Fuzzy Image Processing and Applications with MATLAB®

Fuzzy Image Processing and Applications with MATLAB®

Tamalika Chaira
Ajoy Kumar Ray

CRC Press
Taylor & Francis Group
Boca Raton London New York

CRC Press is an imprint of the
Taylor & Francis Group, an **informa** business

CRC Press
Taylor & Francis Group
6000 Broken Sound Parkway NW, Suite 300
Boca Raton, FL 33487-2742

International Standard Book Number: 978-1-4398-0708-8 (Hardback)

Library of Congress Cataloging-in-Publication Data

Chaira, Tamalika.
Fuzzy image processing and applications with MATLAB / authors, Tamalika Chaira, Ajoy Kumar Ray.
p. cm.
Includes bibliographical references and index.
ISBN 978-1-4398-0708-8 (hardcover : alk. paper)
1. Image processing--Digital techniques--Mathematics. 2. Fuzzy mathematics. 3. MATLAB. I. Ray, Ajoy K., 1954- II. Title.

TA1637.C4774 2010
621.36'7015113223--dc22 2009032787

Visit the Taylor & Francis Web site at
http://www.taylorandfrancis.com

and the CRC Press Web site at
http://www.crcpress.com

Contents

Preface

During the last two decades, the art and science of image processing have witnessed significant developments and have found applications in many active areas, such as remote sensing, medical imaging, video surveillance, and so on. In contrast to classical image-analysis techniques that use crisp mathematics, fuzzy set theoretic techniques provide an elegant foundation and a set of rich methodologies in diverse image-processing tasks. In view of this, it has become extremely important to present this science along with some applications in a textbook.

Fuzzy image processing—Application in MATLAB® is an exciting and dynamic branch of image processing, which has received lots of importance during the last decade. Ever since the introduction of fuzzy set theory by Professor L.A. Zadeh in 1965, there has been an explosion of interest and a significant growth of active application of fuzzy set theory in image processing in such areas as enhancement, segmentation, filtering, edge detection, content-based image retrieval, pattern recognition, and clustering. The progress in this field can be seen through the introduction of the increasing numbers of software products in the market.

There are many texts available in the market that deal with the fundamentals of image processing and its applications using crisp sets. In this book, we have attempted to introduce the concepts of fuzzy set theory and their applications in image processing. Imprecision arises in image processing in several of the following ways: (1) ambiguity arising from imprecision in the grey level of the image; (2) imprecision in the geometry of the object; (3) imprecision in the definition of the edges or the boundary of the objects of an image; and (4) uncertainty in the knowledge representation, object recognition, and image interpretation.

Keeping in mind the above considerations, we felt the need to write this book, which deals with the application of fuzzy set theory to various image-processing operations, such as thresholding, segmentation, edge detection, enhancement, clustering, color retrieval, and so on.

The results of the experimental results using fuzzy set theory are presented in each chapter. Finally, a concise summary highlighting the ideas discussed in each chapter is included.

There are many excellent textbooks on image processing. An important feature of this book, however, is the inclusion of a separate chapter in which a brief introduction to MATLAB and its implementations has been presented.

This book is useful for undergraduate and graduate students in universities world wide. It is extremely useful for teachers, scientists, engineers, and all those who are interested in the fuzzy set theory of image processing.

Organization of This Book

This book contains 11 chapters. Each chapter begins with an introductory text, covers the theory, and then culminates with the applications of fuzzy image processing. To understand the techniques of fuzzy image processing, it is essential to have a background knowledge of the principles of fuzzy sets. Chapter 1 introduces the fundamentals of fuzzy subsets and their operations. Some of the essential mathematics on fuzzy sets are explained in this chapter, along with a number of examples. Chapter 2 deals with some of the basic concepts of fuzzy image processing. Various applications of fuzzy image processing are also discussed in this chapter. Chapter 3 covers different types of similarity and distance measures and various types of fuzzy entropy measures. Fuzzy preprocessing is covered in detail in Chapter 4. This chapter includes different methods of fuzzy image enhancement that increase the contrast of the image. Various fuzzy–classical and fuzzy–fuzzy filters for the removal of impulse/Gaussian noise are discussed in this chapter. Chapter 5 covers fuzzy thresholding techniques. It contains diverse thresholding and fuzzy thresholding schemes that use different types of membership functions. Experimental results on different types of images are included in this chapter. Chapter 6 introduces fuzzy match–based region extraction. Given a model image and a scene image, the model image is allowed to search in the scene image to find the closest pattern in the scene image. Results of fuzzy match–based segmentation on some intensity and texture images are presented in this chapter. Chapter 7 deals with fuzzy edge detection, where the edges of an image are detected to preserve the structural information of the image. Different types of fuzzy methods are explained along with examples of real time images. Chapter 8 provides a detailed account of fuzzy retrieval of color images. Given a query image and an image database, images in the database that are most similar to the query image are retrieved. Various approaches to color retrieval, which use fuzzy histogram and fuzzy similarity measures, are discussed. Some methods on fuzzy retrieval are also discussed and the retrieval results are displayed.

Chapter 9 discusses the fundamentals of pattern classification, which includes various supervised and unsupervised classification schemes. The limitations of statistical classifiers are identified, followed by a discussion on how fuzzy techniques in pattern classification will be more appropriate in many applications. Fuzzy supervised classification, fuzzy clustering, and fuzzy similitude–based classifiers are elaborated in this chapter. Chapter 10 discusses the application of fuzzy set theory in remotely sensed images. Various types fuzzy neural network, fuzzy multilayer perceptron, fuzzy back propagation classifier are discussed. Also, fuzzy counter propagation network in classification of remote sensed images is highlighted in this chapter.

Chapter 11 introduces MATLAB and includes the programs for each application in image processing that are mentioned in this book. Finally, problems for each chapter are included at the end of the book. To help the reader to acquire practical knowledge, a set of projects have been suggested, which are also included at the end of the book.

During the writing of this book, we have received the support and encouragement of many individuals. We gratefully acknowledge the contributions of each one of those who have helped in various ways in the completion of this project. Professor G.S. Sanyal, former director of IIT Kharagpur; Professor Arun Kumar Majumdar, Computer Science and Engineering Department, IIT Kharagpur; and Dr. Tinku Acharya have always been a great source of inspiration and encouragement throughout our research. Many others of the Computer Vision Laboratory of IIT Kharagpur have helped us in shaping this book. Dr. C.V. Jawahar, IIIT Hyderabad; Dr. B.M. Mannan, Philips Innovation; Professor K.M. Bhurchandi, SRKNEC, Nagpur; Professor Sukesh Kumar, BIT Mesra; Rajesh Thakur, KITS, Bhubaneswar; Koushik Mallick; Arumoy Mukhopadhyaya; and J. Chakraborty of the Electronics and ECE Department have all helped us in various ways. We acknowledge their contributions.

We are thankful to Dr. Chandan Chakraborty, Roshan Joy Martis, and Rusha Patra of the School of Medical Science and Technology, IIT Kharagpur, for meticulously going through some of the chapters of this book.

Finally, this book would not have been complete without the support of our near and dear ones.

Tamalika Chaira
Ajoy Kumar Ray

Authors

Tamalika Chaira, PhD received her bachelor's degree in electronics and communication from Bihar Institute of Technology, Sindri, India; her master's degree in electronics and communication from BE College, Shibpur; and her PhD in image processing from the Indian Institute of Technology, Kharagpur, India, in 1993, 2000, and 2003, respectively. Her research interests include image processing, fuzzy logic, intuitionistic fuzzy logic, and medical information processing. She has the following achievements to her credit: 1 Intel U.S. patent, 10 papers in international journals, several Springer-Verlag book chapters, and several papers in international conferences. She is listed in the International Biographic Centre, Cambridge, United Kingdom, and *Marquis Who's Who in Science and Engineering*, United States. She is a reviewer of several Elsevier and IEEE journals. Dr. Chaira is also a member of Soft Computing in Image Processing. Currently, she is a young scientist in the Department of Science and Technology working at the Centre for Biomedical Engineering, Indian Institute of Technology, Delhi, India.

Professor Ajoy Kumar Ray is currently the vice-chancellor of Bengal Engineering and Science University, Howrah, India. He received his BE in electronics and telecommunication engineering from Bengal Engineering College, Sibpur, India, and his MTech and PhD from the Indian Institute of Technology (IIT), Kharagpur, India. Prior to this, he was the head of the School of Medical Science and Technology and a professor of electronics and electrical communication engineering at IIT Kharagpur.

Professor Ray has successfully completed 17 research projects as principal investigator, sponsored by Intel Corporation, Texas Instruments, in addition to those funded by agencies such as Defense Research and Development Organization (DRDO), Department of Science and Technology (DST), the Department of Atomic Energy, and the Department of Information Technology, India. He was at the University of Southampton during 1989–1990 and was the head of the research division at Avisere Inc., United States, during 2004–2005. He has coauthored four books published by international publishing houses. He is the coinventor of six U.S. patents filed jointly with Intel Corporation, as well as three patents filed jointly with Texas Instruments. He has coauthored more than 90 research papers in international journals and conferences.

As secretary and chairman of the Nehru Museum of Science and Technology, Professor Ray has conceptualized and created several galleries on science and technology. His research interests include image processing, machine intelligence, soft computing, and molecular imaging in disease detection.

1

Fuzzy Subsets and Operations

1.1 Introduction

In classical set theory, a set consists of several elements and the elements may or may not present in the set. The membership value of each element in a classical crisp set is either 0 or 1, depending on its absence or its presence in the set. Such a set does not support any fractional membership and each element in it has either full membership or zero membership.

In 1965, Professor L.A. Zadeh introduced fuzzy set theory on the basis of the principles of uncertainty, ambiguity, and vagueness. He suggested that the classes of objects encountered in the real world do not always have precisely defined membership values. There are sets that do not have a rigid demarcating boundary and thus there is a gradual transition from zero to unity membership. These sets are known as fuzzy sets and the elements in these sets have different membership values in the interval [0, 1]. Fuzzy sets have been extensively used in many application areas such as image processing, pattern recognition, decision-support systems, and so on, and they have been effectively used to model the uncertainties, imprecisions, and vaguenesses inherent in these application areas.

1.2 Concept of Fuzzy Subsets and Membership Function

In classical set theory, a set is defined as a collection of elements having a certain property, each of which belongs to the set. So the characteristic function takes either the value of 0 or 1.

Let us assume a classical set, X, called the universe, whose elements are denoted as x, that is, $X=\{x_1, x_2, \ldots, x_n\}$. Also consider a subset A of the set X such that an element x of X is a member of A if

$$\mu_A(x) = 1$$

Otherwise, for all other elements of X that are not members of A, we can write $\mu_A(x)=0$, where $\mu_A(x)$ is termed as the characteristic or membership function of the elements in the set.

Example 1.1

Consider the following set

$X = \{x_1, x_2, x_3, x_4, x_5, x_6, x_7\}$ and its subset A, where

$A = \{x_1, x_2, x_3, x_6\}$

Subset A can be represented after incorporating the characteristic function as

$$\mu_A(x_1) = 1, \quad \mu_A(x_2) = 1, \quad \mu_A(x_3) = 1, \quad \mu_A(x_6) = 1.$$

Thus, A can be written as

$$A = \{(x_1,1), (x_2,1), (x_3,1), (x_4,0), (x_5,0), (x_6,1), (x_7,0)\}$$

or

$$A = \{x, \mu_A(x), x \in X\}$$

Now let us imagine a situation where the characteristic or membership function of the elements in a set can take any value from 0 to 1. This means that each element in the set has a fractional membership, depending on the extent of its presence in the set (such as partially, moderately, or fully present). This membership function, defined on A, assumes a characteristic value $\mu_A(x_i) = 1$ for those elements x_i that fully belong to set A, and for those elements that do not belong to set A, $\mu_A(x_i) = 0$. The elements having partial membership in the set have membership values $0 < \mu_A(x_i) < 1$. This membership concept can also be represented as

$$A = \{(x_1/0.8), (x_2/0.7), (x_3/0.9), (x_4/0.4), (x_5/0.3), (x_6/1), (x_7/0.3)\}$$

where x_i is an element of the set A, followed by the membership value of the element x_i, which lies between 0 and 1, with 0 signifying "no membership" and 1 signifying "full membership." It is a measure of the degree of belongingness of an object in a set.

1.2.1 Membership Function

In a fuzzy set, the degree of membership of an element signifies the extent to which the element belongs to a fuzzy set, that is, there is a gradation of membership value of each element in a set. The membership functions can be viewed as mappings of diverse human choices to an interval [0, 1]. Thus, a fuzzy set is a more generalized set where the membership values lie between 0 and 1.

Example 1.2

Let us take an example of a set of persons and "age." A fuzzy subset OLD is defined as "to what degree is person x old?" To each person in the universe of discourse, we have to assign a degree of membership in the fuzzy subset OLD. The easiest way to deal with this is by using a membership function based on the person's age.

$$\begin{aligned} \text{Old }(x) &= 0, \quad \text{if age}(x) \le 50 \\ &= (\text{age}(x) - 50)/20 \quad \text{if } 50 < \text{age}(x) \le 70 \\ &= 1 \quad \text{if age}(x) > 70 \end{aligned}$$

This has been shown graphically in Figure 1.1.

There are many different ways in which one can represent the membership function [1,5,9,10]. Some of these are trapezoidal, triangular, exponential, Gaussian, Gamma, S-membership function, and so on. All the information contained in a fuzzy set is described by its membership function. These functions can either be (a) monotonic or (b) non-monotonic.

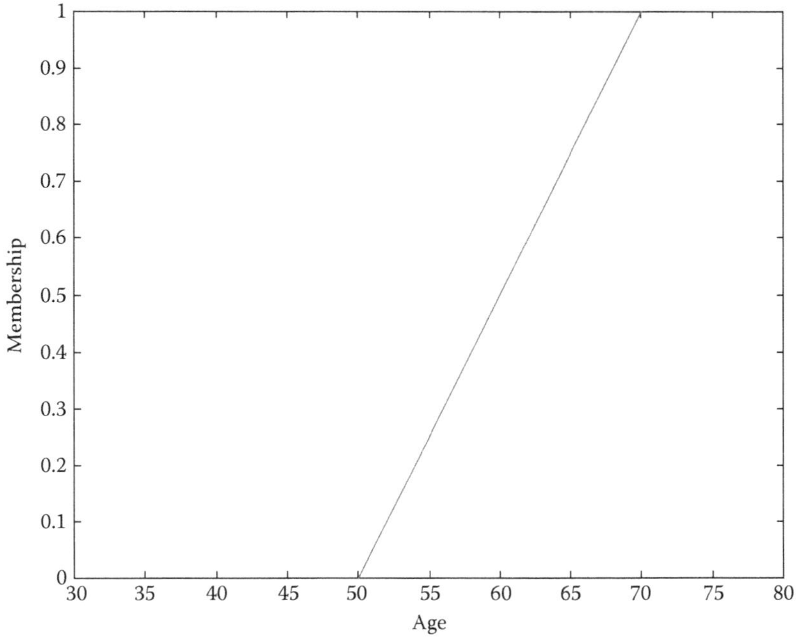

FIGURE 1.1
Membership function of "age."

Example 1.3

A set of "bright pixels" in an image is a monotonic concept, where the membership function of the fuzzy set "bright" increases monotonically with the pixel gray value x. On the other hand, there are fuzzy sets like "moderate-intensity pixels," where the membership function is essentially non-monotonic. We present in the following some of the most commonly used monotonic and non-monotonic families of membership functions.

1. Non-monotonic functions
   a. *Triangular membership function*

$$\begin{aligned}\mu(x) &= 0 \quad \text{if } x \le a \\ &= \frac{x-a}{b-a} \quad \text{if } a \le x \le b \\ &= \frac{c-x}{c-b} \quad \text{if } b \le x \le c \\ &= 0 \quad c \le x\end{aligned}$$

   a, b, and c, are the three parameters as shown in Figure 1.2. a and c are at the feet of the triangle and b is the vertex of the triangle.
   b. *Trapezoidal membership function*

$$\begin{aligned}\mu(x) &= 0 \quad \text{if } x \le a \\ &= \frac{x-a}{b-a} \quad \text{if } a \le x \le b \\ &= 1 \quad \text{if } b \le x \le c \\ &= \frac{d-x}{d-c} \quad \text{if } c \le x \le d \\ &= 0 \quad d \le x\end{aligned}$$

   a, b, c, and d are the four parameters as shown in Figure 1.3. a and d are at the feet of the trapezium, while b and c lie at the shoulder of the trapezium. It is to be noted that if $b=c$, then the trapezoid becomes a triangle.

2. Monotonic functions
   a. *Zadeh's S-membership function* [12]

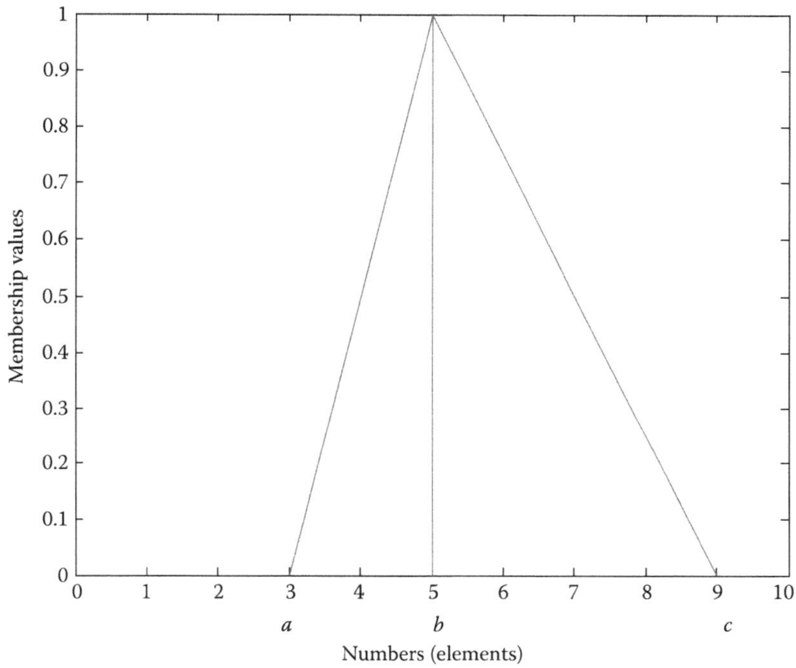

FIGURE 1.2
Triangular membership function.

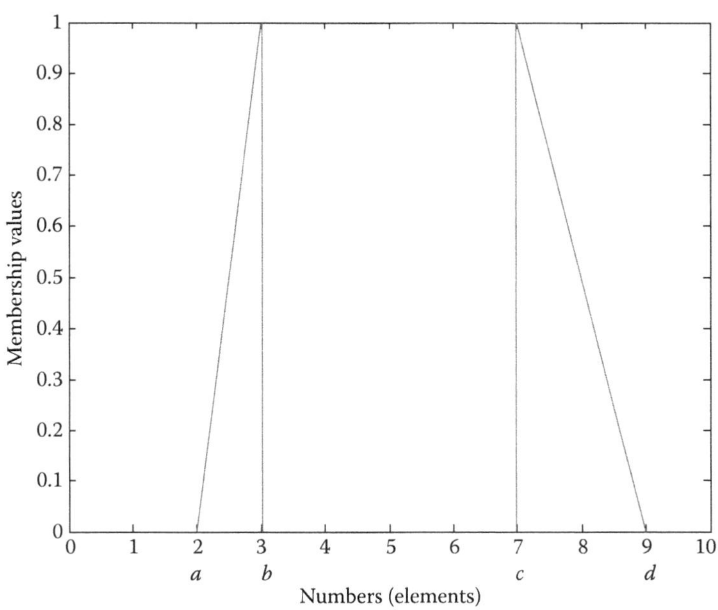

FIGURE 1.3
Trapezoidal membership function.

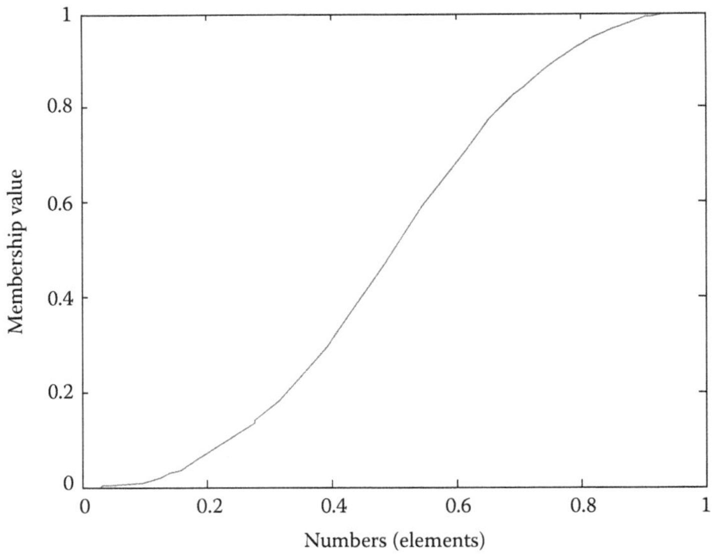

FIGURE 1.4
S-membership function.

$$
\begin{aligned}
\mu(x) &= 0 \quad x < a \\
&= 2[(x-a)/(c-a)]^2 \quad a < x < b \\
&= 1 - 2[(x-c)/(c-a)]^2 \quad b < x < c \\
&= 1 \quad x > c
\end{aligned}
$$

b is any value between a and c.

For $a = x_{min}$, $c = x_{max}$, the membership function plot is shown in Figure 1.4 for a normalized set.

b. *Gaussian membership function* (Figure 1.5)

$$\mu(x) = \exp\left(-\frac{(x-m)^2}{2 \cdot \sigma^2}\right), \quad m = \text{mean},$$

σ is the standard deviation.

c. *Inverse membership function* (Figure 1.6) [5]

$$\mu(x) = \frac{1}{(1 + k \cdot x)}, \quad k > 0$$

k is a constant and can take any value, but the value of k is normally taken as the inverse of the difference between a maximum and minimum values in a set.

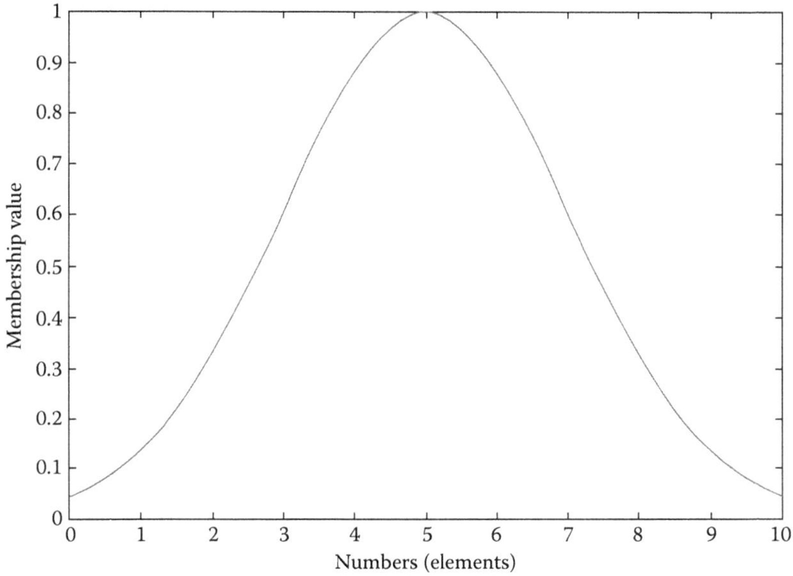

FIGURE 1.5
Gaussian membership function.

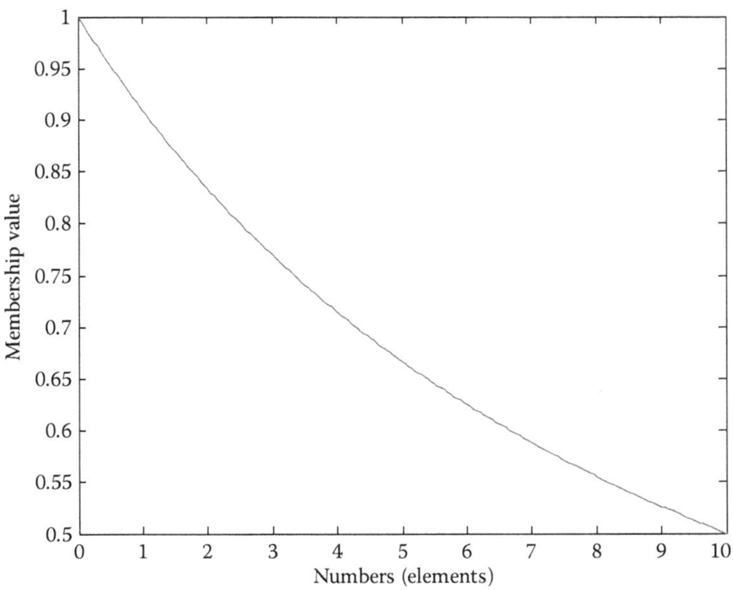

FIGURE 1.6
Inverse membership function.

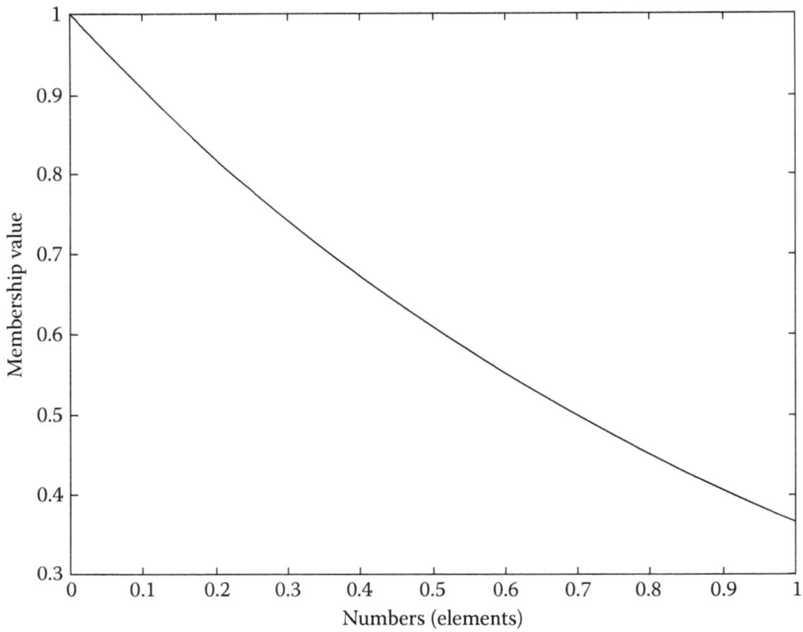

FIGURE 1.7
Exponential membership function.

d. *Exponential membership function* (Figure 1.7)

$$\mu(x) = e^{-kx}, \quad k > 0$$

k is a constant and can be any value.

In the example, k assumes a value as the inverse of the difference of the maximum and minimum values in a set.

e. *Gamma membership function* [1]

$$\mu(x) = \exp(-c.|x-b|), \quad b \text{ is a location parameter.}$$

c is a constant and $c > 0$. c can take the value which is the inverse of the difference of the maximum and minimum values in a set.

There are many other well-known membership functions estimation techniques suggested by different authors and these are stated as follows:

1. Strategies involving similarity and distance measure by Biswas and Majumdar [2] and Zadeh [12].
2. Rank ordering of prototypes according to the compatibility of the features in the set by Dutta Majumder and Chaudhuri [3].

3. Scaling the membership function using theorems of representation and uniqueness on an interval scale by Norwich and Türksen [8].
4. Estimating membership degree of an object using the relative preference method by Saaty [10]. This is applicable in cases where the necessary initial information of the pair-wise comparison is available from experts.

Some of the techniques suggest the estimation of membership function by the addition of several fuzzy variables with known membership grades by Diskant [4]. The essence of each of the above-mentioned approaches lies in the fact that most of these methods do not require a large number of prototypes with known statistics for estimating the membership function. Moreover, the estimation of the membership function should be chosen in such a way as to be consistent with the specification of the data set.

A few terms associated with membership function are explained below:

1. Core—The core of a membership function of a fuzzy set A is that region which is characterized by a complete membership in a fuzzy set A.

   The core thus consists only of those elements whose membership values $\mu_A(x) = 1$.
2. Support—The support of a membership function of a fuzzy set A is that region which is characterized by a nonzero membership. The support thus consists only of those elements whose membership values are greater than 0, $\mu_A(x) > 0$.
3. Boundary—The boundary of a membership function is that region which is characterized by elements possessing nonzero membership but not complete membership, that is, $0 < \mu_A(x) < 1$.
4. Alpha cut (α-cut) of a fuzzy set A with $\mu_A(x)$ as the membership function, defined on X for a given α is given as

$$= \{x/\mu_A(x) \geq \alpha\}$$

5. Strong alpha cut of a fuzzy set A with $\mu_A(x)$ as the membership function, defined on X for a given α is given as

$$= \{x/\mu_A(x) > \alpha\}$$

6. Level set of a fuzzy set is the set of all levels α that represent distinct α-cuts of the fuzzy set A. It can be noted here the 1-cut of A (all the elements that have membership = 1) is called the core of A.
7. Height of A is the largest membership grade obtained by any element in A.

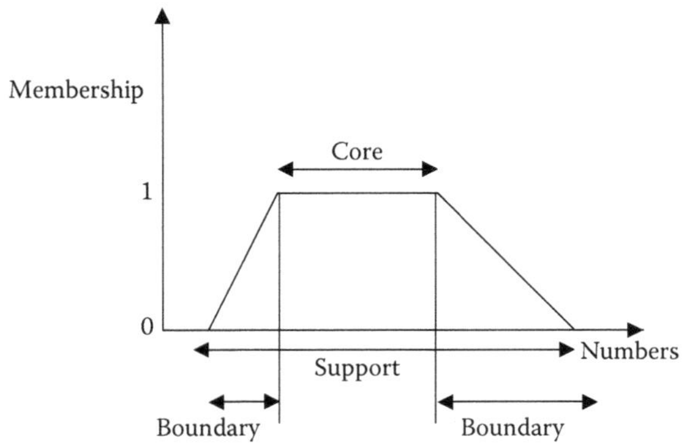

FIGURE 1.8
Core, support, boundary of a fuzzy set.

8. A fuzzy set is called *normal* if its height is 1, that is, if there is at least one point with $\mu_A(x)=1$. Otherwise, it is called *subnormal*. Figure 1.8 shows the core, support, and the boundary of a fuzzy set.

The concepts discussed above can be exemplified using an example.

Example 1.4

Let us take a set of natural numbers

$$A = \{2/0, 3/0.2, 4/0.5, 6/0.7, 7/0.8, 9/0.9, 10/1.0\}$$

1. Core—{10/1.0}
2. Support—{3/0.2, 4/0.5, 6/0.7, 7/0.8, 9/0.9, 10/1.0}
3. 0.4-cut of set A={4/0.5, 6/0.7, 7/0.8, 9/0/9, 10/1.0}
4. 0.8-cut of set A={7/0.8, 9/0.9, 10/1.0}
5. Level set of A(0.2-cut)={0.2, 0.5, 0.7, 0.8, 0.9, 1.0}
6. The height of the above set is 1 since the maximum membership of an element in this set is 1.
7. The set A is "normal" since its height is "1"

The **crossover point** in a fuzzy set A is an element whose membership value is 0.5.

1.3 Linguistic Hedges

In representing the human concepts, we use many linguistic terms, such as bright, moderately bright, and very bright, in our everyday lives, which do not carry any precise meaning in a quantitative sense in conventional mathematics. As fuzzy mathematics deals with uncertainty and vagueness, these terms can be modeled using fuzzy mathematics.

Also quite often, the user needs to modify the membership function depending on his or her requirements. When a fuzzy element with a certain membership function is accompanied with any of these linguistic variables, its membership function also gets modified. Thus, the hedge transforms a fuzzy set to a new fuzzy set by modifying the shape of the membership function.

In fuzzy reasoning, there is a need to intensify the fuzzy set with some linguistic terms, such as very, extremely, more or less, or a complement. There are also other requirements in the image where the regions are approximated as close, near, or somewhat. Some of these modifiers that explain the linguistic terms are as follows:

Very: This modifier can be referred to as concentration, since it intensifies the membership. When this is operated on the membership functions, the membership values are modified as $\mu_{new}=\mu^2$

More or less, somewhat, or rather: It dilutes the membership value of the elements of a fuzzy set and is defined as $\mu_{new} = \sqrt{\mu}$.

Very-very: This modifier can be referred to as extreme and is mathematically denoted as $\mu_{new}=\mu^4$.

Intensify: This modifier is mathematically denoted as $\mu_{new}=2\mu^2$ if $\mu \leq 0.5$ and $\mu_{new}=1-2(1-\mu)^2$ if $\mu > 0.5$.

Plus: This modifier is mathematically denoted as $\mu_{new}=\mu^{1.25}$.

Minus: This modifier is mathematically denoted as $\mu_{new}=\mu^{0.75}$.

Not: This is a complement-type modifier and is mathematically denoted as $\mu_{new}=1-\mu$.

The effect of applying the linguistic terms on a triangular membership function of random numbers is shown in Figure 1.9.

1.4 Operations on Fuzzy Sets

This section provides an introduction to basic operations on fuzzy sets, such as union, intersection, and negation.

Union: Membership function of the union of two fuzzy sets, $X=\{x/\mu_X(x)\}$ and $Y=\{y/\mu_Y(y)\}$ with membership functions μ_X and μ_Y, respectively, is defined as the maximum of the two membership functions. This is called the maximum criterion.

$$\mu_{X\cup Y} = \max(\mu_X, \mu_Y)$$

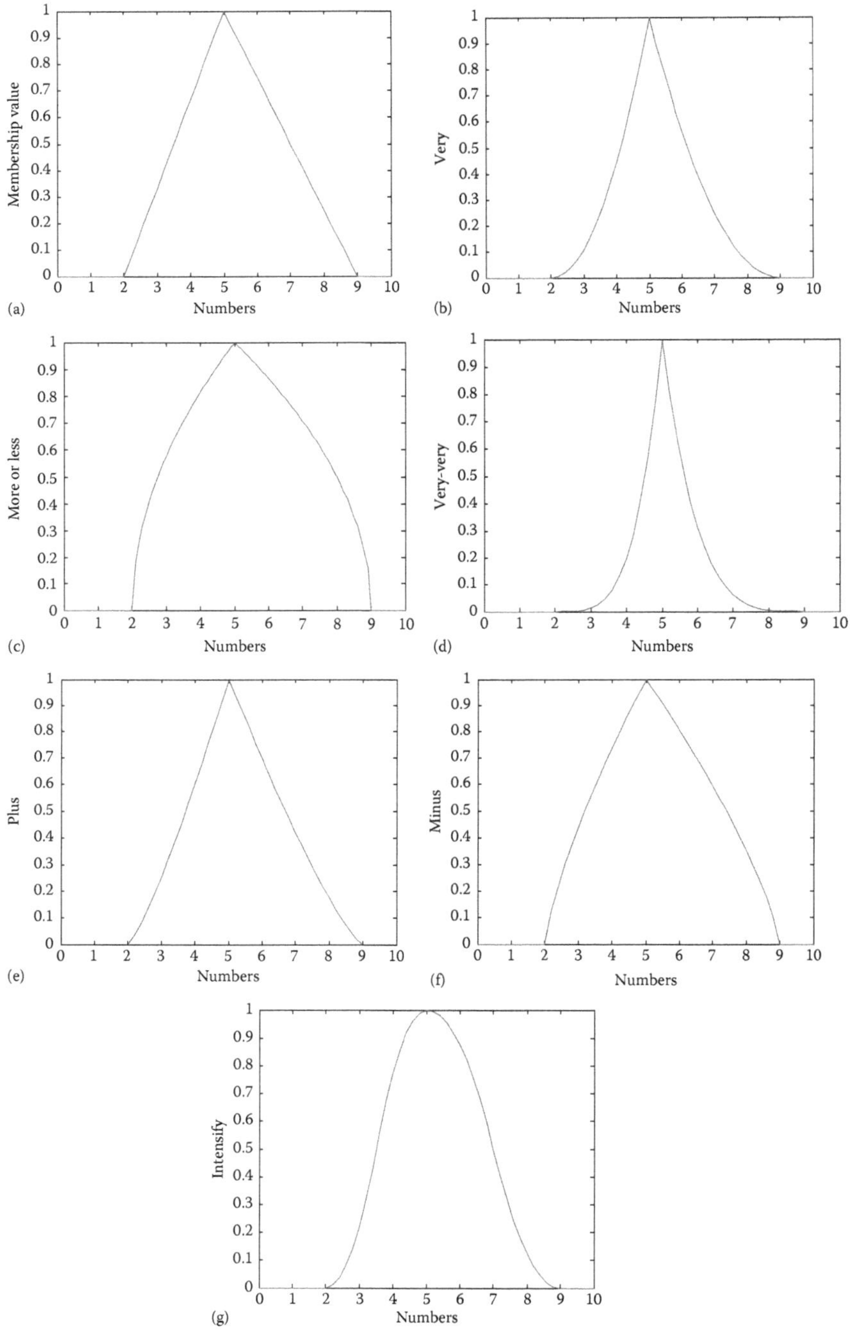

FIGURE 1.9
Modifiers: (a) triangular membership function (existing), (b) "very" modifier, (c) "more or less" modifier (d) "very-very" modifier, (e) plus modifier, (f) minus modifier, and (g) intensify modifier.

The union operation in fuzzy set theory is equivalent to the OR operation in Boolean algebra.

Intersection: Membership function of the intersection of two fuzzy sets, $X=\{x/\mu_X(x)\}$ and $Y=\{y/\mu_Y(y)\}$ with membership functions μ_X and μ_Y, respectively, is defined as the minimum of the two individual membership functions. This is called the minimum criterion.

$$\mu_{X \cap Y} = \min(\mu_X, \mu_Y)$$

The intersection operation in fuzzy set theory is equivalent to the **AND** operation in Boolean algebra.

Complement: Membership function of the complement of a fuzzy set $X=(x/\mu_X(x))$ with membership function μ_X is defined as the negation of the membership function.

$$\mu_{\overline{X}} = 1 - \mu_X$$

The complement operation in fuzzy set theory is equivalent to the NOT operation in Boolean algebra.

Algebraic product: The algebraic product of two fuzzy sets, $X=\{x/\mu_X(x)\}$ and $Y=\{y/\mu_Y(y)\}$, with membership functions μ_X and μ_Y is given as $\mu_{X.Y}(z)=\mu_x(x)\mu_Y(y)$.

Algebraic sum: The algebraic sum of two fuzzy sets, $X=\{x/\mu_X(x)\}$ and $Y=\{y/\mu_Y(y)\}$, with membership functions μ_X and μ_Y is given as $\mu_{X+Y}(z)=\mu_X(x)+\mu_Y(y)-\mu_X(x)\cdot\mu_Y(y)$.

Power: Power of a fuzzy set $X=\{x/\mu_X(x)\}$ is defined as $\mu_X{}^{\alpha} = \mu_X(x)^{\alpha}$.

Difference: The difference of two fuzzy sets, $X=\{x/\mu_X(x)\}$ and $Y=\{y/\mu_Y(y)\}$, with membership functions μ_X and μ_Y is defined as $X-Y=X\cap\overline{Y}$.

Equality: The equality between two fuzzy sets, $X=\{x/\mu_X(x)\}$ and $Y=\{y/\mu_Y(y)\}$, with membership functions μ_X and μ_Y is defined as $X=Y \Leftrightarrow \mu_X(x)=\mu_Y(y)$.

Empty set: Empty set or the null set ϕ of a fuzzy set X is that set where the membership values of all the elements are zero.

Inclusion: Inclusion relation between two fuzzy sets, $X=\{x/\mu_X(x)\}$ and $Y=\{y/\mu_Y(y)\}$, is defined point wise, that is, $X \subseteq Y \Leftrightarrow \mu_X(x) \le \mu_Y(y)$.

Example 1.5

Consider two fuzzy sets X and Y with the following membership functions:

$$X = \{(x_1/0.4), (x_2/0.7), (x_3/0.8), (x_4, 1)\}$$

$$Y = \{(x_1/0.3), (x_2/0.6), (x_3/0.5), (x_4, 0.9)\}$$

The results of the various operations discussed above are shown below.

1. Union

$$\text{Since } 0.3 < 0.4, 0.6 < 0.7, 0.5 < 0.8, 0.9 < 1,$$

$$Y \subset X$$

$$X \cup Y = \{(x_1/0.4),(x_2/0.7),(x_3/0.8),(x_4,1)\}$$

$$X \cap Y = \{(x_1/0.3),(x_2/0.6),(x_3/0.5),(x_4,0.9)\}$$

$$\bar{Y} = \{(x_1/0.7),(x_2/0.4),(x_3/0.5),(x_4,0.1)\}$$

$$X \cdot Y = \{(x_1/0.12),(x_2/0.42),(x_3/0.4),(x_4,0.9)\}$$

$$X + Y = \{(x_1/0.58),(x_2/0.88),(x_3/0.9),(x_4,1)\}$$

$$X - Y = X \cap \bar{Y} = \{(x_1/0.4),(x_2/0.4),(x_3/0.5),(x_4,0.1)\}.$$

1.5 Fuzzy Relations

As with ordinary relations, fuzzy relations have also been studied by a number of authors, such as Zadeh [11,13], Kaufmann [6], Klir [7], and Zimmerman [12]. Suppose that P is a product set and μ is the membership function; the fuzzy relation is then a subset of the product set P taking its value of μ. As in the theory of ordinary sets, fuzzy relations are fuzzy subsets of $X \times Y$.

In the relation $\mathfrak{R} = \{(x, y), \mu_R(x, y)\}, \forall (x, y) \in X \times Y, \forall x \in X, \forall y \in Y$

$$\text{Let} \quad X = \{x_1, x_2, x_3, x_4)$$

$$Y = \{y_1, y_2)$$

The product set $X \times Y$ contains a set of ordered pairs:

$$X \times Y = \{(x_1, y_1),(x_1, y_2),(x_2, y_1),(x_2, y_2),(x_3, y_1),(x_3, y_2),(x_4, y_1),(x_4, y_2)\}$$

$\mu_R(x, y)$ is the values of the ordered pairs.

If the values of the ordered pairs are $\mu(x_1, y_1) = 0.1$, $\mu(x_1, y_2) = 0.4$, $\mu(x_2, y_1) = 0.5$, $\mu(x_2, y_2) = 0.9$, $\mu(x_3, y_1) = 0.5$, $\mu(x_3, y_2) = 0.2$, $\mu(x_4, y_1) = 0$, $\mu(x_4, y_2) = 1.0$, the above values can be written as

$\Re$	y_1	y_2
x_1	0.1	0.4
x_2	0.5	0.9
x_3	0.5	0.2
x_4	0	1.0

FIGURE 1.10
Fuzzy relation.

$$\Re = \{((x_1, y_1)/0.1), ((x_1, y_2), 0.4), ((x_2, y_1), 0.5), ((x_2, y_2), 0.9), ((x_3, y_1), 0.5), ((x_3, y_2), 0.2), ((x_4, y_1), 0), ((x_4, y_2), 1.0)\}$$

Fuzzy relation $\Re$ can be represented in a matrix form as shown in Figure 1.10.

$\mu_{\Re}(x, y)$ is the value of the ordered pair in the product set $X \times Y$ in unit interval [0, 1], that is, each ordered pair assumes a degree of membership value in the interval [0, 1]. Fuzzy relations are the fuzzy subsets in product spaces. Just as in fuzzy sets, fuzzy relations also have similar operations.

Example 1.6

Let us consider two fuzzy relations $\Re_1$ and $\Re_2$ in Figure 1.11a and b and the operations on the two relations are shown in Figure 1.11c–g.

If $\Re_1$ and $\Re_2$ are two relations in the same product space, fuzzy operations are defined as

The union $\mu_{\Re_1 \cup \Re_2}(x, y) = \max\{\mu_{\Re_1}(x, y), \mu_{\Re_2}(x, y)\}, \quad (x, y) \in X \times Y$

(a)

$\Re_1$	y_1	y_2
x_1	0.3	0.3
x_2	0.5	0.4
x_3	0.0	0.2
x_4	0.0	1.0

(b)

$\Re_2$	y_1	y_2
x_1	0.4	0.9
x_2	0.5	0.6
x_3	0.1	0.4
x_4	0.3	1.0

(c)

$\Re_1 \cap \Re_2$	y_1	y_2
x_1	0.3	0.3
x_2	0.5	0.4
x_3	0.0	0.2
x_4	0.0	1.0

(d)

$\Re_1 \cup \Re_2$	y_1	y_2
x_1	0.4	0.9
x_2	0.5	0.6
x_3	0.1	0.4
x_4	0.3	1.0

(e)

$\Re_1 \cdot \Re_2$	y_1	y_2
x_1	0.12	0.27
x_2	0.25	0.24
x_3	0.0	0.08
x_4	0.0	1.0

(f)

$\Re_1 + \Re_2$	y_1	y_2
x_1	0.58	0.93
x_2	0.75	0.76
x_3	0.1	0.52
x_4	0.3	1.0

(g)

$\overline{\Re_1}$	y_1	y_2
x_1	0.7	0.7
x_2	0.5	0.6
x_3	1.0	0.8
x_4	1.0	0.0

FIGURE 1.11
(a) Relation $\Re_1$, (b) relation $\Re_2$, (c) intersection, (d) union, (e) algebraic product, (f) algebraic sum, and (g) complement of operations on fuzzy relations.

The intersection $\mu_{\Re_1 \cap \Re_2}(x,y) = \min\{\mu_{\Re_1}(x,y), \mu_{\Re_2}(x,y)\}, \quad (x,y) \in X \times Y$

The algebraic product of two fuzzy relations $\mu_{\Re_1 . \Re_2}(x,y) = \mu_{\Re_1}(x,y) \cdot \mu_{\Re_2}(x,y)$

The algebraic sum of two relations $\mu_{\Re_1 + \Re_2}(x,y) = \mu_{\Re_1}(x,y) + \mu_{\Re_2}(x,y) - \mu_{\Re_1}(x,y) \cdot \mu_{\Re_2}(x,y)$

The complement of a relation $\mu_{\overline{\Re_1}}(x,y) = 1 - \mu_{\Re_1}(x,y)$

1.5.1 Composition of Two Fuzzy Relations

Fuzzy relations in different product spaces can be combined with each other using the operator *composition*, to form a max–min composition [6, 7]. For any two fuzzy relations $\Re_1$ and $\Re_2$ such that

$$\Re_1 \subset X \times Y$$

$$\Re_2 \subset Y \times Z$$

the max–min composition is denoted as

$$\Re_1 \circ \Re_2(x,z) = \max_y(\min(\mu_{\Re_1}(x,y), \mu_{\Re_2}(y,z)))$$

where $x \in X$, $y \in Y$, $z \in Z$.

Example 1.7

An example with two fuzzy relations $\Re_1(x, y)$ and $\Re_2(y, z)$ is shown in Figure 1.12a and b.

$$(x_1, z_1) = \min(\mu_{\Re_1}(x_1, y_1), \mu_{\Re_2}(y_1, z_1))$$

$$= \min (0.2, 0.8)$$

$$= 0.2$$

$$\min(\mu_{\Re_1}(x_1, y_2), \mu_{\Re_2}(y_2, z_1)) = \min(0.3, 0.1) = 0.1$$

$$\min(\mu_{\Re_1}(x_1, y_3), \mu_{\Re_2}(y_3, z_1)) = \min(0, 0.6) = 0$$

$$\min(\mu_{\Re_1}(x_1, y_4), \mu_{\Re_2}(y_4, z_1)) = \min(1, 0.4) = 0.4$$

$\Re_1$	y_1	y_2	y_3	y_4	y_5
x_1	0.2	0.3	0	1	0.5
x_2	0.3	0.5	0	0.1	1
x_3	0.6	0	0.9	0.3	0.4

(a)

$\Re_2$	z_1	z_2	z_3
y_1	0.8	0	0.3
y_2	0.1	0.8	1
y_3	0.6	0.3	0.7
y_4	0.4	0.3	0
y_5	0.3	1	0.2.

(b)

$\Re_1 \circ \Re_2$	z_1	z_2	z_3
x_1	0.4	0.5	0.3
x_2	0.3	1	0.5
x_3	0.6	0.4	0.7

(c)

FIGURE 1.12
(a) Relation $\Re_1$, (b) relation $\Re_2$, and (c) composition of two fuzzy relations.

$$\min\left(\mu_{\Re_1}(x_1, y_5), \mu_{\Re_2}(y_5, z_1)\right) = \min(0.5, 0.3) = 0.3$$

$$\Re_1 \circ \Re_2 (x_1, z_1) = \left\{(x_1, z_1), \max(0.2, 0.1, 0, 0.4, 0.3)\right\}$$
$$= \left\{(x_1, z_1), 0.4\right\}$$

In this way, for x_i, z_j, $i=1, 2, 3$ and $j=1, 2, 3$, the (x_i, z_j) values are computed. The resulting composition is shown in Figure 1.12c.

If $\Re$ were to be a relation in the space $Y\times Y$, that is, $\Re \subset Y\times Y$, then it can be written as

$$\Re \circ \Re = \Re^2$$

From this, it can be written in the general form as $\Re \circ \Re \circ \ldots = \Re^k$.

1.5.2 Fuzzy Binary Relation

A subset of $A\times A$ is called a binary relation. Fuzzy binary relations have been presented by many authors such as Kaufmann [6], Klir and Yuan [7], and so on.

When $x, y \in A$ and the values of ordered pair lie in the interval [0, 1], then a fuzzy binary relation exists in $A\times A$. Consider $A=[x_1, x_2, x_3]$, and $\Re$ is a set of ordered pairs in $A\times A$ having values $\mu_\Re$ then the matrix in Figure 1.13 is a fuzzy relation in $A\times A$.

$\Re$	x_1	x_2	x_3
x_1	0.7	0.1	0.8
x_2	0.1	1	0.0
x_3	0.8	0.0	0.7

FIGURE 1.13
Fuzzy binary relation.

Properties of fuzzy binary relation:

1. Symmetric property

   A relation is said to be symmetric if $y \Re x$ holds whenever $x \Re y$ for every $x, y \in A$. A symmetric fuzzy binary relation is defined as $\mu_{\Re}(x, y) = \mu_{\Re}(y, x), \forall(x, y) \in A \times A$. If relation $\Re$ is a set of real numbers and $x, y \in R$, then the relation in Figure 1.14 is an example of a symmetric relation.

2. Reflexive property

   A relation is said to be reflexive if $x \; \Re \; x$ holds for every $x \in A$. It is defined as $\forall(x, x) \in A \times A$, $\mu_{\Re}(x, x) = 1$. Figure 1.15 is an example of a reflexive relation.

3. Transitive property

   A relation is said to be transitive if $x \; \Re \; z$ whenever $x \; \Re \; y$ and $y \; \Re \; z$ holds for every $x, y, z \in A$.

$\Re$	x_1	x_2	x_3
x_1	0.7	0.1	0.8
x_2	0.1	1	0.0
x_3	0.8	0.0	0.7

FIGURE 1.14
Symmetric property.

$\Re$	x_1	x_2	x_3
x_1	1	0.1	0.7
x_2	0.1	1	0.3
x_3	0.8	0.0	1

FIGURE 1.15
Reflexive property.

If $x, y, z \in A$ and $\forall(x, y), (y, z), (x, z) \in A \times A$, then the max–min transitivity is defined as

$$\mu_{\Re}(x, z) \geq \max_{y}\left[\min\left(\mu_{\Re}(x, y), \mu_{\Re}(y, z)\right)\right]$$

This transitivity can be presented as

$$\Re \circ \Re \subset \Re$$

Transitivity can also be defined in another way as max-product transitivity. This is defined as

$$\mu_{\Re}(x, z) \geq \max_{y}\left[\mu_{\Re}(x, y) \cdot \mu_{\Re}(y, z)\right]$$

An example of a fuzzy transitive relation is "*A* is much larger than *B*."

1.5.3 Transitive Closure of Fuzzy Binary Relation

Transitive closure is the smallest fuzzy relation that is transitive and is contained in $\Re$. The elements of transitive closure have the smallest possible membership grades.

The transitive closure of a fuzzy binary relation is

$$\widehat{\Re} = \Re \cup \Re^2 \cup \Re^3 \cup \ldots$$

The transitive closure of any fuzzy binary relation is a transitive binary relation, that is, $\widehat{\Re}^2 \subset \widehat{\Re}$. This denotes that $\Re$ is transitive and helps in constructing a transitive relation from any relation.

0.2	0.1	0.4
0.1	1	0.1
0.3	0.0	0.7

(a)

0.3	0.1	0.4
0.1	1	0.1
0.3	0.1	0.7

(b)

0.3	0.1	0.4
0.1	1	0.1
0.3	0.1	0.7

(c)

0.3	0.1	0.4
0.1	1	0.1
0.3	0.1	0.7

(d)

0.3	0.1	0.4
0.1	1	0.1
0.3	0.1	0.7

(e)

FIGURE 1.16
Transitivity and transitive closure. (a) $\Re$, (b) $\Re^2$, (c) $\Re^3$, (d) $\hat{\Re} = \Re \cup \Re^2$, and (e) $\hat{\Re} \circ \hat{\Re} = \hat{\Re}^2$.

Example 1.8: Example for Transitivity and Transitive Closure

Consider a fuzzy relation $\Re$ in Figure 1.16a. It is observed from Figure 1.16b that $\Re^2 \supset \Re$ instead of $\Re^2 \subset \Re$. It is not following the transitive property.

Now we shall check for transitive closure, that is, $\hat{\Re} \subset \Re \cup \Re^2 \cup \Re^3 \cup \ldots$

$\Re^2$ is calculated and then $\Re^3$. We see that $\Re^2 = \Re^3$ in Figure 1.16c, so further calculation of $\Re^4$ is not required.

$\hat{\Re} = \Re \cup \Re^2$ is computed and is given in Figure 1.16d and simultaneously $\hat{\Re}^2$ is computed as given in Figure 1.16e.

It is observed that $\hat{\Re}^2 \subset \hat{\Re}$.

This proves that the relation $\Re$ is transitive.

If a relation satisfies all the three properties, that is, reflexive, symmetry, and transitive, then the relation is a fuzzy similitude or a fuzzy equivalence relation.

1.6 Summary

This chapter deals with fuzzy subsets, fuzzy membership functions, fuzzy operations, and fuzzy relations. Fuzzy set deals with the uncertainty in the element present in a set. This uncertainty is represented in the form of a membership function. There are different types of membership functions, namely, trapezoidal, triangular, Gaussian, exponential, gamma, inverse, and S membership functions.

Similar to the operations and relations on crisp sets, there are also operations and relations on fuzzy sets. These operations are union, intersection, complement, negate, etc. Fuzzy relations such as reflexivity, symmetricity, transitivity, and equivalence relation are also discussed along with examples.

References

1. Chaira, T. and Ray, A.K., Segmentation using fuzzy divergence, *Pattern Recognition Letters*, 12(24), 1837–1844, 2003.
2. Biswas, T. and Majumdar, A.K., A multistage fuzzy classifier for recognition of handprinted characters, *IEEE Transactions on Systems, Man, and Cybernetics, SMC*, 11(12), 834–838, 1981.
3. Dutta Majumder, D. and Chaudhuri, B.B., On membership evaluation in fuzzy sets, in *Fuzzy Information and Decision Process*, M.M. Gupta and E. Sanchoz (Eds.), North-Holland, Amsterdam, the Netherlands, 1982.
4. Diskant, H., About membership function estimation, *Fuzzy Sets and Systems*, 5, 141–147, 1981.
5. Huang, L.K. and Wang, M.J.J., Image thresholding by minimizing the measure of fuzziness, *Pattern Recognition*, 28, 41–51, 1995.
6. Kaufmann, A., *Introduction to the Theory of Fuzzy Subsets: Fundamental Theoretical Elements*, vol. 1, Academic Press, New York, 1980.
7. Klir, G. and Yaun, B., *Fuzzy Set and Fuzzy Logic: Theory and Application*, Prentice Hall India Pvt. Limited, New Delhi, India, 1977.
8. Norwich, A.M. and Türksen, I.B., The construction of membership functions, in *Fuzzy Sets and Possibility Theory*, R.R. Yager (Ed.), Pergamon Press, New York, pp. 61–67, 1982.
9. Pal, S.K. and Rosenfeld, A., Image enhancement and thresholding by optimization of fuzzy compactness, *Pattern Recognition Letters*, 7, 77–86, 1988.
10. Saaty, T.L., Exploring the interface between hierarchies, multiple objectives, and fuzzy sets, *Fuzzy Sets and Systems*, 1(1), 57–68, 1978.
11. Zadeh, L.A., Fuzzy sets, *Information and Control*, 8, 338–353, 1965.
12. Zimmerman, H.J., *Fuzzy Set Theory and Its Application*, Kluwer Academic, Amsterdam, the Netherlands, 1996.
13. Zadeh, L.A., Similarity relations and fuzzy orderings, *Information Science*, 3, 177–200, 1971.

2

Image Processing in an Imprecise Environment

2.1 Introduction

With the emergence of fuzzy logic concepts, its significance and application in digital image processing became evident. Prewitt is possibly the first person to note the scope of incorporating the fuzzy set theoretic concepts in image processing. The application of fuzzy set theory in image processing took formal shape only in the 1980s with the pioneering research carried out by Pal et al. [25–29] and Rosenfeld [33,34].

The pixel values, which constitute an image, may not be precise and there is basically an inherent imprecision or uncertainty embedded in a digital image. While attempting to design automated systems for scene analysis and interpretation, it may be a good idea to consider the fact that a computer vision system is usually embedded with uncertainty and vagueness, which needs to be taken care of appropriately. Appropriate modeling of this imprecision arising in a physical phenomenon is an important task in many applications of image processing.

Fuzzy set theory is thus useful in handling various uncertainties in computer vision and image-processing applications. Fuzzy image processing is a collection of different fuzzy approaches to image processing that can understand, represent, and process the images. It has three main stages, namely, image fuzzification, modification of membership values, and, if necessary, image defuzzification, as shown in Figure 2.1.

Fuzzification and defuzzification are the important steps that are required in processing the images with fuzzy techniques. The main part of fuzzy image processing lies in the membership plane, as shown in Figure 2.2. Membership degree is the degree of belongingness of the pixels in an image. Fuzzification is the transformation of gray level of the image to a membership function using some mathematical function. After fuzzification, depending on the user's requirements, the membership values of the gray level are modified using an appropriate fuzzy technique. Techniques for modifying the membership values of the pixels in an image may involve fuzzy rule–based approach, fuzzy linguistic hedges, fuzzy integration, and so on.

Fuzziness is inherently embedded in nature and is bound to be reflected in its reproduction in the form of images. Let us now see the logical reasons behind the applicability of fuzzy notions in image processing by Prewitt [32]:

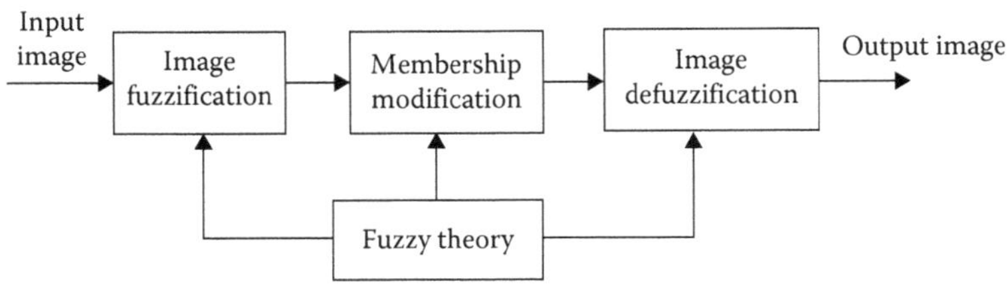

FIGURE 2.1
Structure of fuzzy image processing.

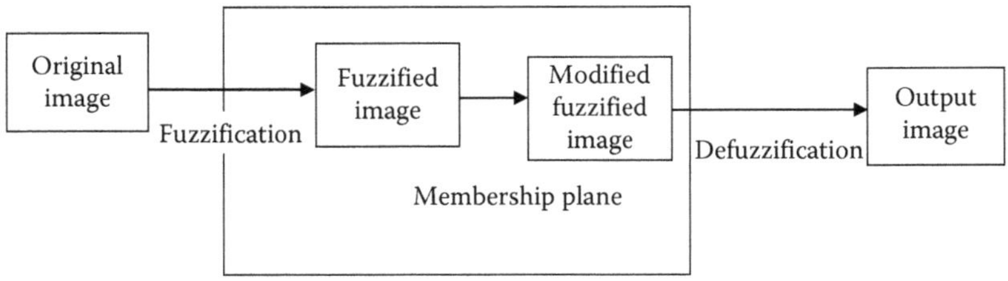

FIGURE 2.2
Steps of fuzzy image processing.

1. Images are two-dimensional (2-D) projections of the objects in a three-dimensional (3-D) world and this mapping is performed by a camera, which can operate in optical, near infrared, or even thermal bands of the electromagnetic spectrum. It can be stressed that some information is lost during the mapping from 3-D space to 2-D images, and thereby uncertainty is introduced into images.
2. In multi-valued gray-level images, uncertainty exists due to the variability of gray value, along with the statistical uncertainty due to the randomness. The pixel values in digital images need not be considered as precise constants, and should be viewed as fuzzy numbers.
3. Many of the image definitions, such as boundaries of objects or the homogeneity of the segments in an image are vague in nature and require fuzzy notions for their characterization.
4. In a similar manner the contrast between the objects and the background can also be viewed as fuzzy in nature.
5. Ambiguity is often present in the final interpretation of the image. Since the human understanding system is never crisp or precise, it is important to incorporate soft decision-making since a hard decision can often prove to be costly.

In crisp image analysis, the image regions, edges, and other features are usually considered to be crisp. But the regions or the edges in an image are not always properly defined and this uncertainty or vagueness should be considered when using fuzzy decision-making. Let us consider an example. In remotely sensed images, rivers often manifest in an imprecise way. When the water level is deep in the middle of a river, the shallowness increases as one moves toward the river bank and the actual banks may be dry and devoid of any water. The delineation of the water body from the bank is often not possible using crisp mathematical techniques. Indeed the two segments cannot be precisely defined by any crisp technique, and thus fuzzy modeling or soft delineation is often preferred.

Uncertainty in an image can be due to the grayness ambiguity and spatial ambiguity. Grayness ambiguity is the uncertainty in deciding whether a pixel is white or black and spatial ambiguity is the vagueness in defining the image object boundaries, edges, etc. Thus, instead of the crisp values that are required to make the image features crisp, values between 0 and 1 are used to make the image regions, boundaries, edges, and other features fuzzy.

Next, when any decision is taken at a certain level, it will also have an impact on the subsequent higher levels. Due to the presence of uncertainties in an image, the system should process the information at each level.

It can also be pointed out here that if the uncertainties in an image, such as image regions and edges, are computed, they will surely yield more information on the scene, compared to a crisp description of it. Fuzzy decision, in this case, is very useful, as one can have more information than the crisp decision. In some applications such as image interpretation and image understanding where expert knowledge is required, fuzzy logic is thus very useful.

2.2 Image as a Fuzzy Set

Let A be an image of size $M \times N$ having L levels and x_{ij} be the gray level at the pixel position (i, j), $i=0, 1, 2, 3, \ldots, M-1, j=0, 1, 2, 3, \ldots, N-1$. Let $\mu_A(x_{ij})$ denote the membership value of the pixels of the image A with $0 \le \mu_A(x_{ij}) \le 1$ where $\mu_A(x_{ij}) = 1$ denotes full membership and $\mu_A(x_{ij}) = 0$ denotes nonmembership. Intermediate degrees are graded accordingly. Membership value represents the information, say, for example, the brightness, of the pixel at position (i, j).

In fuzzy set theory, an image can be represented as

$$\sum_{i=0}^{M-1} \sum_{j=0}^{N-1} \left[x_{ij}, \mu_A\left(x_{ij}\right) \right], \quad \forall x \in A, \, i = 0, 1, 2, \ldots, M-1, \quad j = 0, 1, 2, \ldots, N-1.$$

2.3 Fuzzy Image Processing

Fuzzy image processing is based on fuzzy logic and uses its logical, set theoretical, and relational aspects. The most important frameworks that can be used to construct the foundations of fuzzy image processing are fuzzy geometries, measures of fuzziness/information, measures of fuzzy similarity, fuzzy clustering, fuzzy rule–based approaches, fuzzy mathematical morphologies, and fuzzy grammars.

2.3.1 Foundations of Image Processing

2.3.1.1 Fuzzy Geometry

Geometrical components such as area, perimeter, and compactness in the image components play a key role in image processing. Due to the vague nature of the image boundaries/edges, regions, the images are not always 'crisply' defined rather it is sometimes regarded as fuzzy subsets of the image. Fuzzy digital geometry (Rosenfeld [33]) may be very useful when fuzzy geometrical measures such as fuzzy area, perimeter, compactness, and the index of area coverage may be used to measure the geometrical fuzziness of different regions of an image. The application areas of geometrical fuzziness are feature extraction, image representation, and image segmentation.

As an example, Pal and Rosenfeld [34] used the concept of fuzzy compactness (using area and perimeter) to threshold an image where the membership values of the pixels of an image are computed using the S membership function. Optimum threshold is obtained by minimizing the fuzzy compactness. Chaira and Ray [6] computed the membership values of the pixel using the Gamma membership function and then maximized the fuzzy compactness to obtain the optimum threshold.

2.3.1.2 Measures of Fuzziness/Information

Measure of fuzziness is the information on "how fuzzy the image really is." One type of measure of fuzziness is the index of fuzziness by Kaufmann [19]. Indices of fuzziness is of two types: linear and quadratic indices of fuzziness. It is the distance between a fuzzy set to its nearest ordinary crisp set, that is, 0 or 1. The amount of fuzziness is 0 or minimum if all the membership values are 0 or 1, and is maximum if all the membership values are equal to 0.5. Another measure of fuzziness is the logarithmic fuzzy entropy by Luca and Termini [22], which is measure of fuzziness in a fuzzy set. Besides this, there are many works on measures of image information such as fuzzy entropy, fuzzy divergence, fuzzy similarity measures, fuzzy correlation, and weighted fuzzy expected value [6,12,16,26,28,37]. The main application areas of measures of fuzziness are in image thresholding, segmentation, image enhancement, retrieval, and so on.

As an example, Chaira and Ray [6] computed the membership values of the pixel using the Gamma membership function and then minimized the indices of fuzziness to obtain an optimum threshold. Pal and Rosenfeld [34] minimized the indices of fuzziness to obtain the threshold.

2.3.1.3 Rule-Based Systems

The theory of fuzzy IF–THEN rules was proposed by Zadeh. In the field of image-processing applications, mainly in image recognition and scene analysis, expert knowledge is required. The most common way to represent human knowledge is the use of some verbally formulated rules to form a natural language expression. The general form of a rule is as follows:

IF X is A THEN Y is B. Here A is the premise or the antecedent and B is the conclusion or the consequent.

It means that if the fact is known then the conclusion can easily be derived. They can be applied in various image-processing applications such as image enhancement, image segmentation, edge detection, classification, and so on. If we interpret the image features as linguistic variables, then we can use fuzzy IF–THEN rules to segment and classify the image into different regions or to detect the edges of an image. A simple fuzzy segmentation rule may seem as follows:

IF the pixel is *dark*
AND its neighborhood is also *dark* **AND** *homogeneous*
THEN it *belongs* to the background

2.3.1.4 Fuzzy Clustering

Fuzzy clustering is a useful approach for image classification, segmentation, and retrieval. It is actually a grouping of similar pixels in a single cluster and of different pixels in different clusters. Pixels belonging to a cluster have similar properties while pixels belong to different clusters have different properties. Algorithms such as fuzzy c-means, possibilistic c-means, and mixed clustering [1,3,11,16,25,35,39] can be used to build clusters. Pixels may belong to one or more clusters where each element is associated with membership value depending on the belongingness to each cluster.

2.3.1.5 Fuzzy Mathematical Morphology

Morphology is a nonlinear image-processing technique that deals with the shape of the image features. Sometimes there is a need to obtain the skeleton of an image or the dilation/erosion of an image or any other forms of an image for segmentation, thinning, or object recognition. In such cases, morphological operations are required. The basic image transformation such as dilation, erosion, opening, and closing using set operations like difference, sum, maximum, and minimum of two images are applied for low-level and intermediate image processing [14]. Morphological operations use a structuring element or mask and it interacts with the image to see how it fits or misses the shapes in the image. The sizes and shapes of the structuring element are related to the geometric properties of the image objects. The structuring element is placed at all the locations of an image to obtain a user-required transformation of an image. By modifying the weights of the masks, it is possible to tune morphological operators to specific image-processing tasks. But the shapes of the objects in an image in this real world are not always properly or crisply defined, and in that case fuzzy set theory is very useful that deals with vagueness and ambiguity. As the mathematical morphological operators are set like operations, so fuzzy set theory is easily applied to mathematical morphology. Initially fuzzification is done using any membership function and then basic morphological operators are applied on it. Fuzzy morphology is applied in image thresholding, enhancement, edge detection, and object recognition [5,15,24].

2.3.1.6 Fuzzy Grammars

The concept of grammar was introduced by linguistics for obtaining the structural descriptions of sentences of a language rather than numerical values. So, it can be more appropriate to apply the formal language theory by Fu [13]. But the real patterns are often ill-defined and the formal grammar becomes very rigid. Fuzzy language can be a better tool for defining ill-defined structural information [29,30,44]. A fuzzy grammar produces a language that is a fuzzy set of strings and each string has a membership value that denotes the degree of belongingness of that string to the language. The grade of membership of an unknown pattern is obtained by the max–min rule.

2.4 Some Applications of Fuzzy Set Theory in Image Processing

Fuzzy image analysis finds applications in many computer-vision systems including robot vision, object recognition, biomedical image processing, remotely sensed scene analysis, and so on. However, applying a single image analysis method may not always yield reliable results. It is observed that when an ensemble of image-processing techniques are applied to an image, appropriate results are likely to be obtained. In the following, a few applications of fuzzy set theory in image processing are discussed.

1. *Image contrast enhancement:* The object of image enhancement is to transform the original image into another image that is more suitable for further processing steps. It increases the contrast of the image by

making the dark pixels darker and the bright pixels brighter. One of the most popular conventional enhancement methods is histogram hyperbolization, equalization [2,32]. But this method may not always yield good results, since the pixel gray levels in an image are imprecise. Many researchers proposed different fuzzy enhancement methods such as histogram hyperbolization by Tizoosh [41], contrast intensification, and fuzzy expected value by Schneider [38] and so on by using fuzzy techniques. In many cases, global histogram approaches to enhancement fail to achieve satisfactory results as these methods consider the whole image in its totality. Locally adaptive histogram equalization (LAHE) techniques often increase the performance of enhancement procedure in such cases. In such a case, an $n \times m$ neighborhood window is selected, which is moved at each pixel location and the enhancement is performed locally until the whole image is scanned.

2. *Image segmentation:* Segmentation is one of the most important steps to analyze the image. It partitions the image into disjoint sets corresponding to objects in the scene. There are various types of segmentations, such as thresholding, edge detection, and region-based segmentation, and they are performed depending on the requirement of the user. Each region in a segmented image possesses homogeneous properties with respect to features, such as gray level, texture, or color and the property is different for different regions in an image. In many cases the region boundaries may not have a sharp transition and so a crisp decision as to whether the pixel belongs to a region is difficult to ascertain. In such cases, fuzzy set theory is very useful as it incorporates the uncertainties of the objects, the region definitions, and the features in the form of membership function. Fuzzy set theory and fuzzy logic are then introduced within traditional image-segmentation techniques, for example, in fuzzy thresholding, clustering, supervised segmentation, and rule-based techniques [10,20,23,25,36,37,40].

3. *Boundary detection and representation:* Boundary detection is an important process in image processing. It carries the structural information thus reducing drastically the data to be processed. The boundaries indicate the location and the shape of the objects. Jain et al. [17] state that "An edge point locates a pixel where there is significant local intensity change; an edge fragment is a collection of edge points; an edge detector produces either a set of edge points or edge fragments." Thus edge is present when there is a change in gray-level intensity or gray level. Edge images can be termed as gradient images. There are real time images such as medical images, remotely sensed images, and others, where the edges are not distinct due to poor contrast and unequal illumination. The crisp decision, in such cases, whether an edge is present in the image is very difficult. Fuzzy method is

very useful in alleviating such a type of problem. There are many fuzzy edge detection techniques such as fuzzy template based edge detector [7,8,21], fuzzy entropy based edge detection, and fuzzy reasoning–based methods, to name but a few [31]. Extensive survey on fuzzy based medical image segmentation and edge detection is given by Bezdek et al. [4].

4. *Image retrieval:* Content-based retrieval of image databases usually involves comparing a query object with the objects stored in the image database, when a query object is provided. In conventional image retrieval systems, images are typically characterized by a set of features such as color, texture, and shape. The query image is searched in the image database based on the extracted features, and then the images in the database are ranked according to the similarity index. But, there may be uncertainty in imprecision of pixel color values and human perception. This uncertainty can be alleviated by using fuzzy set theory and thus the image is characterized by a set of fuzzy features, namely, color, texture, and shape [43]. Fuzzy feature is defined by a membership value that corresponds to the belongingness of the feature vector in the region. One method of image retrieval is the use of fuzzy histogram [18,42], which fuzzifies the histogram of an image using a membership function, formed by considering an image gray level to be around 'n' i.e., $n+1$ or $n-1$. Then fuzzy similarity measures [9] are used to retrieve the query image from the database image.

2.5 Summary

In this chapter, we have attempted to discuss the reasons why a digital image should be viewed as inherently fuzzy in nature. Several applications have been highlighted and some of these have been discussed in greater detail in other chapters of this book.

References

1. Bezdek, J.C. and Pal, S.K., *Fuzzy Models in Pattern Recognition*, IEEE Press, New York, 1992.
2. Banks, S., *Signal Processing, Image Processing and Pattern Recognition*, Prentice-Hall International, Cambridge, U.K., 1990.
3. Bensaid, A., Hall, L.O., Bezdek, J.C., and Clarke, L.P., Partially supervised clustering for image segmentation, *Pattern Recognition*, 29(5), 859–871, 1996.

4. Bezdek, J.C. et al., Medical image analysis with fuzzy models, *Statistical Methods in Medical Research*, 6, 191–214, 1997.
5. Bloch, I. and Maitre, H., Fuzzy mathematical morphologies: A comparative study, *Pattern Recognition*, 28(9), 1341–1387, 1995.
6. Chaira, T. and Ray, A.K, Threshold selection using fuzzy set theory, *Pattern Recognition Letters*, 25, 865–874, 2004.
7. Chaira, T., De, S., and Salvetti, O., A new measure using intuitionistic fuzzy set using Hausdorff metric and its application to edge detection, *The International Journal of Fuzzy Mathematics* 15(3), 645–654, 2007.
8. Chaira, T. and Ray, A.K., A new measure using intuitionistic fuzzy set theory and its application to edge detection, *Applied Soft Computing* (2), 919–927, 2008.
9. Chaira, T. and Ray, A.K., Fuzzy measures for color image retrieval, *Fuzzy Sets and Systems*, 150(3), 545–560, 2005.
10. Clark, M. et al., Automatic tumor segmentation using knowledge-based technique, *IEEE Transaction on Medical Imaging*, 17(2), 187–201, 1998.
11. Delgado, M. et al., Hierarchical clustering to validate fuzzy clustering, in *Proceedings of the IEEE International Conference on Fuzzy Systems*, Yokohama, Japan, pp. 1807–1812, 1995.
12. Fan, J. and Xie, W., Fuzzy divergence and induced fuzzy entropy, *Fuzzy Sets and Systems* 104, pp. 305–314, 1999.
13. Fu, S., *Syntactic Pattern Recognition and Application*, Prentice-Hall, Englewood Cliffs, NJ, 1982.
14. Gonzalez, R.C. and Winz, P., *Digital Image Processing*, Addison-Wesley, Reading, MA, 1977.
15. Grobert, S., Koppen, M., and Nickolay, B., A new approach to fuzzy morphology based on fuzzy integral and its application to image processing, in *Proceedings of the ICPR*, vol. II, Vienna, Austria, pp. 625–630, 1996.
16. Gustafson, E.E. and Kessel, W., Fuzzy clustering with fuzzy covariance matrix, in *Proceedings of the IEEE Conference on Decision and Control*, San Diego, CA, pp. 761–766, 1979.
17. Jain, R. et al., *Machine Vision*, McGraw Hill, New York, 1995.
18. Jawahar, C.V. and Ray, A.K., Fuzzy statistics of digital image, *IEEE Signal Processing Letters*, 3(8), 225–227, 1996.
19. Kaufmann, A., *Introduction to the Theory of Fuzzy Subsets*, Addison-Wesley, New York.
20. Keller, J.M. and Chen, Z., Learning in fuzzy neural networks utilizing additive hybrid operators, in *Proceedings of the International Conference on Fuzzy Logic and Neural Networks*, Iizuka, Japan, pp. 85–87, 1992.
21. Kenneth, H.L., Ho, K.H.L., and Ohnishi, N., FEDGE—Fuzzy edge detection by fuzzy categorization and classification of edges, in *Fuzzy Logic in Artificial Intelligence*, Springer, Berlin/Heidelberg, pp. 182–196, 1995.
22. Luca, D. and Termini, A., Entropy and energy measures of fuzzy sets, in *Advances in Fuzzy Sets Theory and Applications*, M.M. Gupta, R.K. Ragadr, and R.R. Yager (Eds.), Elsevier North Holland, Amsterdam, the Netherlands, pp. 321–328, 1979.
23. Miyajima, K. and Norita, T., Region extraction for real image based on fuzzy reasoning, in *Proceedings of the FUZZ-IEEE'92*, San Diego, CA, pp. 229–236, 1992.
24. Nakatsuyama, M., Fuzzy mathematical morphology to image processing, in *Proceedings of the ANZIIS*, Perth, Western Australia, pp. 75–78, 1993.

25. Pal, N.R., Pal, S.K., and Bezdek, J.C., A mixed C-means clustering model, in *Proceedings of the FUZZ-IEEE-97*, Barcelona, Spain, pp. 11–21, 1997.
26. Pal, S.K., King, R.A., and Hishim, A., Automatic gray level thresholding through index of fuzziness and entropy, *Pattern Recognition Letters*, 1, 141–146, 1983.
27. Pal, S.K. and Ghosh, A. Index of area coverage of fuzzy image subsets and object extraction, *Pattern Recognition Letters*, 11, 831–841, 1990.
28. Pal, S.K. and Pal, N.R., Entropy: A new definition and its application, *IEEE Transactions on Systems, Man and Cybernetics SMC*, 21(5), 1260–1270, 1991.
29. Pal, S.K. and Majumder, D., *Fuzzy Mathematical Approach to Pattern Recognition*, John Wiley & Sons, New York, 1986.
30. Pal, S.K. and Pathak, A., Fuzzy grammars in syntactic recognition of skeleton maturity of X-rays, *IEEE Transactions on Systems, Man and Cybernetics SMC*, 16(5), 657–667, 1986.
31. Perez-Oramas, O., Fuzzy reasoning edge detection operator, in *IAR*, Estrasburgo Francia, pp. 27–32, 1999.
32. Prewitt, J.C., Object enhancement and extraction, in *Picture Processing and Psychopictirics*, B.S. Lipkin and A. Rosenfeld (Eds.), Academic Press, New York, pp. 75–149, 1970.
33. Rosenfeld, A., The fuzzy geometry of image subsets, *Pattern Recognition Letters*, 2, 311–317, 1984.
34. Rosenfeld, A. and Pal, S.K., Image enhancement and thresholding by optimization of fuzzy compactness, *Pattern Recognition Letters*, 7, 77–86, 1984.
35. Russ, J.C., *Image Processing Handbook*, CRC Press, Boca Raton, FL, 1994.
36. Russo, F., A new class of fuzzy operators for fuzzy image processing: Design and implementation, in *Proceedings of the FUZZ-IEEE'93*, vol. 2, San Francisco, CA, pp. 815–820, 1993.
37. Sahoo, P.K., Soltani, S., Wang, A.K., and Chen, Y.C., A survey on image thresholding techniques, *Computer Vision Graphics and Image Processing*, 41(2), 233–260, 1988.
38. Schneider, M. and Craig, M., On the use of fuzzy sets in histogram equalization, *Fuzzy Sets and Systems*, 45, 271–278, 1992.
39. Sobrevilla, P., Fuzzy sets in computer vision, *Mathware and Soft Computing*, 10, 71–83, 2003.
40. Tizhoosh, H.R., *Fuzzy Image Processing (in German)*, Springer, Berlin, 1997.
41. Tizhoosh, H.R., Fuzzy histogram hyperbolization for image enhancement, in *Proceedings of the EUFIT-95*, vol. 3, Aachen, Germany, 1995.
42. Vertan, C. and Boujemaa, N., Use of fuzzy histogram and distance for color image retrieval, Challenges of Image Retrieval, Brighton, pp. 1–6, 2000.
43. Yixing, C. and Wang, J.Z., Looking beyond region boundaries: Region based image retrieval using fuzzy feature matching, http://wang.ist.psu.edu/IMAGE/
44. Ross, T.J., *Fuzzy Logic with Engineering Application*, Wiley & Sons, New York, 2004.

3

Fuzzy Similarity Measure, Measure of Fuzziness, and Entropy

3.1 Introduction

In many practical applications of computer vision, there arises a need

1. To compute the similarity between two objects
2. To classify similar groups or elements in a scene

In such cases, the computation of distance or similarity measures becomes essential to distinguish the similarity between two sets. These measures are very common in many applications that deal with matching-related problems. Suppose $\{a_1 ,a_2, a_3, \ldots, a_n)$ and $\{b_1, b_2, b_3, \ldots, b_n)$ are two sets containing n elements. To find the similarity between the two sets and hence the match value, fuzzy similarity, distance measure, measures of fuzziness, and entropy are calculated between fuzzy sets and fuzzy elements. These fuzzy measures are useful when we deal with fuzzy data. In fuzzy database and information retrieval, a query data is allowed to match with the data in the database and if the matching degree, that is, the similarity measure value, exceeds a prespecified threshold, then the data is retrieved.

In many image-processing tasks, it is often required to compare two images in terms of image features. The following text lists out some examples.

Example 3.1: Region Extractions

Region extraction can be termed as match-based segmentations. Given a query or model image and a large scene image, the regions that are similar to the query image are extracted using some similarity or distance measures. Matching is performed between the model image and the different regions in the scene image. It is applicable in remote sensing where the model image can be agricultural regions, mountainous regions, or other regions, and the scene image can be an image of a landscape. The different types of regions are extracted in the landscape image.

Example 3.2: Image Retrieval

Image retrieval is the retrieving of a query image in the database image using a set of features such as color, texture, and shape. For each image, the features are computed and the features of the query image are compared with the features in each of the images in the database using fuzzy similarity or distance measures. The images that are almost similar to the query images are retrieved.

Example 3.3: Edge Detection

Edges denote the structural information or the boundary of an image. Edges may present in many directions and they are detected using a set of edge-detecting templates that signify the possible directions of the edges. These templates are placed over the image and the similarity is calculated to find the edges of the image.

Example 3.4: Image Thresholding

Thresholding is binarizing the image in two levels as object and background. Optimum threshold is obtained by maximizing or minimizing the entropy or any measures of fuzziness or fuzzy similarity.

Example 3.5: Clustering

Clustering is a grouping of similar pixels in one group and dissimilar pixels in another group. This grouping is done using some similarity or distance measures that are required to the control and the formation of clusters. Clustering may be hard or fuzzy. In hard clustering, pixels are divided into distinct clusters, where each pixel belongs to exactly one cluster. In fuzzy clustering, pixels may belong to one or more clusters where each element is associated with a membership degree. Clustering is very useful in medical imaging where the tumor or calcification regions in ultrasound images are detected.

3.2 Fuzzy Similarity and Distance Measures

In this section, fuzzy similarity and distance measures are described using fuzzy set theory. Fuzzy similarity and distance measures are studied by many authors [3,4,14,15,17]. Let us consider a universal set X and $F(X)$ be a fuzzy set. Let $P(X)$ be the class of all crisp sets of X. Also consider two fuzzy sets, A and B, such that $A, B \in F(X)$. The properties of similarity measure and distance measure between two fuzzy sets A and B with membership functions $\mu_A(x)$ and $\mu_B(x)$ are given as follows.

Similarity measure
A function $S: F(X)^2 \rightarrow [0,\infty]$ is called a similarity measure if it satisfies the following properties:

1. $S(A, B) = S(B, A), \forall A, B \in F(X)$
2. For three fuzzy sets A, B, C $\forall A, B, C \in F(X)$, if $A \subset B \subset C$ then $S(A, B) \geq S(A, C)$ and $S(B, C) \geq S(A, C)$

3. $S(C, \bar{C}) = 0, \forall C \in P(X)$
4. $S(D, D) = \max_{A, B \in F(X)} S(A, B), \forall D \in F(X)$

Distance measure
A function D: $F(X)^2 \rightarrow [0, \infty]$ is called a distance measure if it satisfies the following properties:

1. $D(A, B) = D(B, A), \forall A, B \in F(X)$
2. For three fuzzy sets A, B, C $\forall A, B, C \in F(X)$, if $A \subset B \subset C$ then $D(A, B) \leq D(A, C)$ and $D(B, C) \leq D(A, C)$.
3. $D(A, A) = 0, \forall A \in F(X)$
4. $D(C, \bar{C}) = \max_{A, B \in F(X)} D(A, B), \forall C \in P(X)$

3.2.1 Examples of Fuzzy Distance Measures

In this section, various types of fuzzy distance measures are explained in brief. Consider two fuzzy sets A and B in a finite set $X = \{x_1, x_2, x_3, \ldots, x_n)$ with $\mu_A(x)$ and $\mu_B(x)$ as the membership functions of sets A and B, respectively. The distance measures are as follows:

Hamming distance: The Hamming distance between two fuzzy sets A and B is given as

$$d(A,B) = \sum_{i=1}^{n} |\mu_A(x_i) - \mu_B(x_i)| \tag{3.1}$$

Euclidean distance: The Euclidean distance between two fuzzy sets A and B is given as

$$d(A,B) = \sqrt{\sum_{i=1}^{n} (\mu_A(x_i) - \mu_B(x_i))^2} \tag{3.2}$$

3.2.2 Fuzzy Divergence

Fan and Xie [3] proposed fuzzy divergence from fuzzy exponential entropy by using a single row vector.

Consider a fuzzy set with probabilities $(p_0, p_1, \ldots, p_n)$, the exponential entropy can be written as

$$H = \sum_{i=1}^{n} p_i e^{1-p_i}$$

For fuzzy cases, fuzzy entropy of a fuzzy set of size n can be expressed as

$$H(A) = \frac{1}{n(\sqrt{e}-1)} \sum_{i=1}^{n} \left[(\mu_A(x_i) \cdot e^{1-\mu_A(x_i)} + (1-\mu_A(x_i)) \cdot e^{\mu_A(x_i)} - 1 \right], \; i = 1, 2, \ldots, n \tag{3.3}$$

where $\mu_A(x_i)$ is the membership value of the ith element of set A.

For two fuzzy sets A and B, the amount of information of $\mu_A(x_i)$ and $\mu_B(x_i)$ at ith point is given as

$$e^{\mu_A(x_i)} / e^{\mu_B(x_i)} = e^{\mu_A(x_i)-\mu_B(x_i)}$$

Corresponding to fuzzy entropy in Equation 3.3, fuzzy expected information of discrimination in favor of A against B can be given as

$$D_{11}(A,B) = \sum_{i=1}^{n} \left(1/2 - (1-\mu_A(x_i))e^{\mu_A(x_i)-\mu_B(x_i)}\right)$$

In a similar manner, the fuzzy information of the discrimination of complements A' against B' is given as

$$D_{12}(A',B') = \sum_{i=1}^{n} \left(1/2 - \mu_A(x_i)e^{\mu_B(x_i)-\mu_A(x_i)}\right)$$

Hence, the discrimination of A against B is given as

$$\begin{aligned} D_1(A,B) &= D_{11}(A,B) + D_{12}(A',B') \\ &= \sum_{i=1}^{n} \left(1 - (1-\mu_A(x_i))\, e^{\mu_A(x_i)-\mu_B(x_i)} - \mu_A(x_i)e^{\mu_B(x_i)-\mu_A(x_i)}\right) \end{aligned}$$

Similarly, the discrimination of B against A is given as

$$D_2(B,A) = \sum_{i=1}^{n} \left(1 - (1-\mu_B(x_i))\, e^{\mu_B(x_i)-\mu_A(x_i)} - \mu_B(x_i)e^{\mu_A(x_i)-\mu_B(x_i)}\right)$$

Thus, total fuzzy divergence between the fuzzy sets A and B is

$$\begin{aligned} D(A,B) &= D_1(A,B) + D_1(B,A) \\ &= \sum_{i=1}^{n} \begin{pmatrix} 2 - (1-\mu_A(x_i) + \mu_B(x_i))e^{\mu_A(x_i)-\mu_B(x_i)} \\ -(1-\mu_B(x_i) + \mu_A(x_i))e^{(\mu_B(x_i)-\mu_A(x_i))} \end{pmatrix} \end{aligned} \tag{3.4}$$

3.3 Examples of Similarity Measures

A few similarity measures between two fuzzy sets A and B containing n number of elements with membership functions $\mu_A(x)$ and $\mu_B(x)$ are given by Wang [14] as follows:

$$\text{a. } S_1(A,B) = \frac{1}{n}\sum_{i=1}^{n} \frac{\min(\mu_A(x_i), \mu_B(x_i))}{\max(\mu_A(x_i), \mu_B(x_i))}$$

When $\mu_A(x_i) = \mu_B(x_i)$, $S_1(A, B) = 1$ implies that the two sets are exactly similar.

$$\text{b. } S_2(A,B) = \frac{\sum_{i=1}^{n} 1 - |\mu_A(x_i) - \mu_B(x_i)|}{n}$$

When $\mu_A(x_i) = \mu_B(x_i)$, $S_2(A, B) = 1$, implies that the two sets are exactly similar.

$$\text{c. } S_3(A,B) = 1 - \frac{\sum_{i=1}^{n} |\mu_A(x_i) - \mu_B(x_i)|}{\sum_{i=1}^{n} (\mu_A(x_i) + \mu_B(x_i))}$$

Here also, when $\mu_A(x_i) = \mu_B(x_i)$, $S_3(A, B) = 1$ implies that the two sets are exactly similar.

3.3.1 Measure Based on Tversky's Model

As is well known, the human biological neural system is extremely proficient in identifying the similarity between objects or patterns that we come across every moment. However, computer-vision researchers pay very little attention to study similarity measures that account for the human perception and judgment. Quite often the similarity and distance measures that we use do not take into account these human perceptions.

Let us suppose that there exists a model which recognizes an object in actual appearance from its replica; this replica is essentially similar to the original object with some minor variations arising due to noise and other distortions. This is essentially a problem of recognition of two identical objects in conditions of uncertainty. If there exists a model where uncertainty is involved, then the similarity index should be such that it eliminates the uncertainty and recognizes the known object. In such cases, the similarity between sets of features of the object and its distorted version can be appropriately calculated. Thus, the assessment of similarity between different objects is believed to be very important for various recognition processes where human judgment is taken into account.

In 1977, Tversky [13] proposed a feature contrast model where he asserted that similarity should be judged by looking into the common and distinctive features of the two sets. Instead of considering stimuli as points in a metric space, Tversky described them using sets of features. A feature set is a set of logic predicates. Tversky suggested that human beings attend more on the common features in similarity judgement. If a, b are two stimuli, A and B are the sets of features, respectively, and $s(a, b)$ is a similarity between a and b, then Tversky's theory is based on the following assumptions:

$$\textit{Matching}: s(a,b) = F(A \cap B, A - B, B - A)$$

$$\textit{Monotonicity}: s(a,b) > s(a,c)$$

whenever

$$A \cap C \subseteq A \cap B,\ A - B \subseteq A - C,\ B - A \subseteq C - A.$$

A function that satisfies matching and monotonicity is called matching function.

Tversky's measure is intended for binary features only. Tolias [12] proposed a generalized Tversky's index (GTI) as a similarity measure that uses fuzzy predicates. It is defined as

$$\text{GTI}\ (A,B;\alpha,\beta) = \frac{f(A \cap B)}{f(A \cap B) + \alpha \cdot f(A - B) + \beta \cdot f(B - A)} \tag{3.5}$$

where

f is a nonnegative and increasing function
A, B are two sets of predicates on the measurements
$\alpha, \beta \geq 0$

The values of α, β determine the relative importance of the distinctive features in the similarity assessment. When $\alpha \neq \beta$, we get a directional similarity measure. When $\alpha > \beta$, the focus is on the distinctive features of set A. Likewise when $\beta > \alpha$, the focus is on the distinctive features of set B. GTI provides a set theoretic index for similarity assessment based on human perception. It compares the saliency of the common features to the saliency of distinctive features. $A \cap B$ denotes the common features, that is, features that are common to both sets A and B. $A - B$ are the distinctive features that are present in a set A and not in set B. In Equation 3.5 any measure of fuzziness, f, or any t-norm can be used for finding the similarity. Using the min t-norm, Tversky Index-Min is defined as

$$\text{TIM}(A,B;\alpha,\beta)$$
$$= \frac{\sum_{i=1}^{n} \min(\mu_A(x_i),\mu_B(x_i))}{\sum_{i=1}^{n} (\min(\mu_A(x_i),\mu_B(x_i)) + \alpha \cdot \min(\mu_A(x_i), 1-\mu_B(x_i)) + \beta \cdot \min(1-\mu_A(x_i),\mu_B(x_i)))}$$

When the membership values of the elements of the two fuzzy sets are similar, it can be inferred that the common features of the two sets are more and the distinctive features are less, and, likewise, the two fuzzy sets are dissimilar if the common features are less and distinctive features are more.

3.3.2 Similarity of Fuzzy Sets Based on Distance

Similarity of fuzzy sets can also be calculated using distance function. There are many approaches by different authors based on the relation between similarity and distance functions. Two of the relations between two fuzzy sets A and B are expressed as follows:

According to Johaniyak and Kovacs [4] and Koczy and Tikk [6]

$$S(A,B) = \frac{1}{1+D(A,B)}$$

Another distance-based similarity function was proposed by Williams and Steele [15]

$$S(A,B) = e^{-\alpha \cdot D(A,B)}$$

where α is a steepness measure. Normally, the value α is kept at 7 when working with a 1-D set. The distance D, can be the Hamming distance, the Euclidean distance, or any other distance measures.

3.4 Measures of Fuzziness

The measure of fuzziness provides a way to measure the fuzziness of a fuzzy set. Different authors [1,2,7,10,11,18] describe measures of fuzziness in a different manner. Kaufmann [5] suggested indices of fuzziness, Yager [18] defined measure of fuzziness as a distance between fuzzy set and its complement, de Luca and Termini considered a measure of fuzziness as a mapping from power set $P(X)$ to $[0,+\infty]$.

The properties of the measures of fuzziness (I) are

1. $I(A)=0$ or minimum iff $\mu_A(x_i)=0$ or 1, $\forall i$
2. $I(A)=1$ or maximum iff $\mu_A(x_i)=0.5$, $\forall i$
3. $I(A)=I(\overline{A})$
4. $I(A)\geq I(B)$, where B is a sharpened version of A such that

$$\mu_B(x_i)\geq\mu_A(x_i)\quad \text{if}\,\mu_A(x_i)\geq 0.5$$

$$\mu_B(x_i)\leq\mu_A(x_i)\quad \text{if}\,\mu_A(x_i)\leq 0.5$$

3.4.1 Index of Fuzziness

The index of fuzziness denotes the degree of ambiguity or fuzziness present in a set by measuring the distance between the membership values of the fuzzy elements of the fuzzy set A and its nearest ordinary set $\tilde{A}$. The index of fuzziness is defined as

$$I(A)=\left(2/n^k\right)\cdot d\left(A,\tilde{A}\right) \tag{3.6}$$

where $d(A,\tilde{A})$ denotes the distance between fuzzy set A and its nearest ordinary set $\tilde{A}$.

An ordinary set $\tilde{A}$ nearest to the fuzzy set A is defined as

$$\mu_{\tilde{A}}(x_i)=0\quad \text{if}\,\mu_A(x_i)\leq 0.5$$

$$=1\quad \text{if}\,\mu_A(x_i)>0.5$$

There are two types of indices of fuzziness: linear index of fuzziness and quadratic index of fuzziness. These are defined as follows:

a. If $k=1$, the distance: d becomes the Hamming distance.

$$d\left(A,\tilde{A}\right)=\sum_{i=1}^{n}\left|\mu_A(x_i)-\mu_{\tilde{A}}(x_i)\right|$$

$$=\sum_{i=1}^{n}\mu_{A\cap\bar{A}}(x_i)$$

$\mu_{A\cap\bar{A}}(x_i)$ is the intersection or common values of the membership degree of the element x_i at ith point in the fuzzy set A and its complement, n, is the number of elements in a fuzzy set.

The linear index (LI) of fuzziness is rewritten as

$$\text{LI}=\left(\frac{2}{n}\right)\sum_{i=1}^{n}\min(\mu_A(x_i),\ (1-\mu_A(x_i))) \tag{3.7}$$

b. If $k=0.5$, d becomes the Euclidean distance, then the quadratic index (QI) of fuzziness is defined as

$$\text{QI} = \left(\frac{2}{\sqrt{n}}\right)\left(\sum_{i=1}^{n}(\mu_A(x_i) - \mu_{\bar{A}}(x_i))^2\right)^{1/2}$$

$$\text{QI} = \left(\frac{2}{\sqrt{n}}\right)\left(\sum_{i=1}^{n}(\min(\mu_A(x_i),(1-\mu_A(x_i)))^2\right)^{1/2} \tag{3.8}$$

3.4.2 Index of Nonfuzziness

The index of nonfuzziness by Pal and Rosenfeld [9] denotes the amount of nonfuzziness or crispness in $\mu_A(x_i)$ by computing the distance between a fuzzy set, A, from its complement version [9].

$$D(A,\bar{A}) = \left[\sum_{i=1}^{n}\left|\mu_A(x_i) - \mu_{\bar{A}}(x_i)\right|\right], \quad i = 1,2,3,\ldots,n$$

where $\mu_{\bar{A}}(x_i) = 1 - \mu_A(x_i)$

3.4.3 Yager's Measure

Yager [18] suggested that the measure of fuzziness depends on the relationship between the fuzzy set and its complement. It is the measure of a lack of distinction between the fuzzy set A and its complement $\bar{A}$. The distance between the fuzzy set A and its complement $\bar{A}$ is denoted as

$$D_p(A,\bar{A}) = \left[\sum_{i=1}^{n}\left|\mu_A(x_i) - \mu_{\bar{A}}(x_i)\right|^p\right]^{1/p}, \quad i = 1,2,3,\ldots,n \tag{3.9}$$

$$\mu_{\bar{A}}(x_i) = 1 - \mu_A(x_i)$$

The measure of fuzziness is defined as

$$S(A) = 1 - \frac{D_p(A,\bar{A})}{|A|^{1/p}}$$

$p=1$, D_1 is a Hamming distance, and if $p=2$, D_2 is a Euclidean distance.

3.5 Fuzzy Entropy

Fuzzy entropy is a measure of quantity of fuzzy information gained from a fuzzy set. It is analogous to the classical Shannon entropy in information theory, but with a slight variation. Shannon's entropy uses the concept of probabilistic theory and it contains randomness uncertainty whereas fuzzy entropy uses the concept of ambiguity and vagueness uncertainty. This vagueness is defined by a membership function.

3.5.1 Logarithmic Entropy

De Luca and Termini uses the concept of Shannon's probabilistic entropy to define fuzzy entropy. It is the entropy of a fuzzy set that is based on the membership function. For a finite set $X=\{x_1, x_2, x_3, \ldots, x_n\}$, the entropy of a fuzzy set A is

$$E(A)=\frac{1}{n}\sum_{i}\left[-\mu_A(x_i)\log[\mu_A(x_i)]-[1-\mu_A(x_i)]\log[1-\mu_A(x_i)]\right],\quad i=1,2,3,\ldots,n \tag{3.10}$$

where $\mu_A(x_i)$ is the membership degree of the element x_i in A.

3.5.2 Shannon Fuzzy Entropy

Sander [10] suggested the Shannon fuzzy entropy that is analogous to the Shannon entropy. For a finite set $X=\{x_1, x_2, x_3, \ldots, x_n\}$ and F_p, $p\in[0,1]$ denotes the constant fuzzy sets that is $F_p(x_i)=p$ for all $x_i\in X$, $i=1, 2, 3,\ldots, n$, then the power of a fuzzy set $f\in[0,1]^X$ is defined as

$$P(X)=\sum_{i=1}^{n} f(x_i)$$

Considering fuzzy entropy as a mapping from $[0,1]^X\rightarrow[0,\infty]$, Shannon fuzzy entropy is defined as

$$H(f)=-c\sum_{i=1}^{n} f(x_i)\log(f(x_i),\quad c>0$$

with $i=1, 2, 3, \ldots, n$

3.5.3 Total Entropy

De Luca and Termini [1,2] introduced the concept of total entropy. It is the measure of uncertainty of information that contains fuzziness and randomness uncertainties. It computes the relationship between the information in an ordinary set and information in a fuzzy set.

Consider an experiment with 'n' number of events $\{x_1, x_2, x_3, \ldots, x_n\}$ where one and only one event may occur in each trial with respective probabilities $\{p_1, p_2, p_3, \ldots, p_n\}$. On fuzzifying, two types of uncertainties may exists:

1. Uncertainty due to the random nature of experiment which is given by the Shannon's probabilistic entropy as

$$H\{p_1, p_2, p_3, \ldots, p_n\} = -\sum_{i=1}^{n} p_i \log p_i$$

   This gives the average amount of information in making a prediction on the elelment.

2. Uncertainty due to the fuzziness of a fuzzy set, which is due to the difficulty in interpretating of x_i:

$$S(\mu_i) = -\mu_i \log \mu_i - (1 - \mu_i) \log(1 - \mu_i)$$

   where μ_i is the membership value of element x_i.

The statistical average, m of $S(\mu_i)$ is given by

$$m\{\mu, p_1, p_2, p_3, \ldots, p_n\} = \sum_{i=1}^{n} p_i S(\mu_i)$$

   This m gives the average amount of difficulty in taking a decision (0 or 1) on elements.

The total entropy is defined as

$$H_{\text{total}} = H(p_1, p_2, p_3, \ldots, p_n) + m\{\mu, p_1, p_2, p_3, \ldots, p_n\} \tag{3.11}$$

Thus from the above two uncertainties, if fuzziness is removed ($m = 0$), then the total entropy, H_{total}, reduces to Shannon's entropy and if randomness is removed, then the total entropy, $H_{\text{total}} = S(\mu_i)$.

The total entropy is also given by Xie and Bedrosian [16]. Consider a set A containing elements 0 and 1 with probabilities p_0 and p_1. Suppose that due to some reasons, the sharpness in an ordinary set A' is changed and a new set so formed is a fuzzy set A. The membership values of the element are changed from 0 to any value in the interval [0,0.5] and from 1 to any value in the interval [0.5,1]. In this way the ordinary set A' is changed to a fuzzy set A. This fuzzy set has two uncertainties: (a) randomness uncertainty in the ordinary set and (b) fuzziness uncertainty due to the fuzziness in a set. The total entropy of a fuzzy set is defined as:

$$H_{\text{total}} = H(p_0, p_1) + \frac{1}{n}\sum_{i=1}^{n} S(\mu_i) \tag{3.12}$$

The drawback of this entropy is an 'equivalance' is formed between fuzzy and Shannon information. If $p_i = \mu_i$, average amount of information produced by a fuzzy set of n elements is "equivalent" to the average amount of Shannon's information produced by the same number of binary sources, which is meaningless. Again, the sum of fuzzy and probabilistic information is not meaningful when they are conceptually different. Further, if fuzziness is removed, the total entropy reduces to $H(p_0, p_1)$, irrespective of the defuzzification process.

3.5.4 Hybrid Entropy

The drawbacks of total entropy are removed by using hybrid entropy introduced by Pal and Pal [7]. The idea is that if a fuzzy set is a generalization of an ordinary set, the entropy of a fuzzy set may be a generalization of classical entropy. For a digital communication over a noisy channel, Xie and Bedrosian assumed two uncertainties, but Pal and Pal suggested that there is only one uncertainty and it is the difficulty in determining whether the incoming symbol is 0 or 1. This difficulty is due to the probability of generation of 0 and 1 by the source and transfer function which makes them fuzzy. A measure related to these uncertainties is called hybrid entropy.

If p_0 and p_1 are the probabilities of occurances of the symbols 0 and 1, respectively, and μ_i is the membership function of a fuzzy set "symbol close to 1," the average likelihood of interpreting a received symbol as 0 is given by

$$E_0 = \frac{1}{n}\sum_{i=1}^{n} (1 - \mu_i) \cdot e^{\mu_i}$$

and the average likelihood of interpreting a received symbol as 1 is given by

$$E_1 = \frac{1}{n}\sum_{i=1}^{n} \mu_i \cdot e^{1-\mu_i}$$

where E_0 and E_1 are the monotonically increasing functions and denotes the likeliness on the possibility of receiving a symbol as 0 or 1.

Then hybrid entropy of a fuzzy set, considering the possibilistic and probabilistic uncertainties, may be given as

$$H_{\text{hybrid}} = -p_0 \log(E_0) - p_1 \log(E_1)$$

3.6 Geometry of Fuzzy Subsets

In an image analysis, it is sometimes required to measure the geometric properties of regions in an image such as area, perimeter, and compactness, depending on the requirement of the user. But as these regions of an image are not crisply defined, the image is considered as a fuzzy set and fuzzy geometric properties are used in image analysis. The standard approach to image analysis and recognition begins with the image segmentation where the image is segmented into various regions.

Rosenfeld [8,9] introduced the concept of fuzzy geometry of image subsets such as area and perimeter. A fuzzy subset of a set S is a mapping μ from S into [0, 1]. For any $P \in S$, $\mu(P)$ is the degree of membership of P in μ and μ is a characteristic function of the subset.

The area of μ is defined as

$$a(\mu) = \int \mu$$

where the integral is taken over the region in which $\mu > 0$.

If μ is a piecewise constant as in a digital image, then a (μ) is the weighted sum of the areas of the regions where μ has constant values and is weighted by these values.

For a piecewise constant, the perimeter of μ is defined as

$$P(\mu) = \sum_{m,n} \sum_{k} |\mu_m - \mu_n| |A_{mnk}|$$

$m, n = 1, 2, 3, \ldots, r$; and $m < n$; $k = 1, 2, 3, \ldots, r_{mn}$

It is the weighted sum of the length of the arcs A_{mnk} along which the mth and nth regions having constant μ values, μ_m and μ_n, respectively, meet and it is weighted by the absolute difference of these values.

The compactness of μ is defined as

$$\text{Comp}(\mu) = \frac{a(\mu)}{p^2(\mu)} \tag{3.13}$$

3.7 Summary

This chapter describes the definitions, properties, and types of distance measures; it also describes the similarity measures, entropy, hybrid entropy, and the measures of fuzziness. The geometry of fuzzy subsets is also dealt with in this chapter.

References

1. De Luca, A. and Termini, S., Entropy and energy measures of fuzzy sets, in *Advances in Fuzzy Sets Theory and Applications*, M.M. Gupta, R.K. Ragade, and R.R. Yager (Eds.), Elsevier, North Holland, Amsterdam, the Netherlands, pp. 321–328, 1979.
2. De Luca, A. and Termini, S., A definition of non-probabilistic entropy in the setting of fuzzy set theory, *Information and Control*, 20, 301–312, 1972.
3. Fan, J. and Xie, W., Distance measure and induced fuzzy entropy, *Fuzzy Sets and Systems* 104, 305–314, 1999.
4. Johaniyak, Z.C. and Kovacs, S., Distance based similarity measures on fuzzy sets, in *Proceedings of the Third Slovakian-Hungarian Joint Symposium on Applied Machine Intelligence*, SAMI 2005, Herl'any, Slovakia, 2005.
5. Kaufmann, A., *Introduction to the Theory of Fuzzy Subsets: Fundamental Theoretical Elements*, vol. 1, Academic Press, New York, 1980.
6. Koczy, L.L. and Tikk, D., Fuzzy rendszerek, Typotex, Budapest, 2000.
7. Pal, N.R. and Pal, S.K., Higher order fuzzy entropy and hybrid entropy of a set, *Information Sciences*, 61(3), 211–231, 1992.
8. Rosenfeld, A., The fuzzy geometry of image subsets, *Pattern Recognition Letters*, 2, 311–317, 1984.
9. Pal, S.K. and Rosenfeld, A., Image enhancement and thresholding by optimization of fuzzy compactness, *Pattern Recognition Letters*, 7, 77–86, 1988.
10. Sander, W., On measure of fuzziness, *Fuzzy Sets and Systems*, 29, 49–55, 1989.
11. Sharhan, S.A. et al., Fuzzy entropy: A brief survey, in *10th IEEE Conference on Fuzzy Systems*, Melbourne, 2001.
12. Tolias, Y.K. et al., Generalized fuzzy indices for similarity matching, *Fuzzy Sets and Systems*, 120, 255–270, 2001.
13. Tversky, A. Features of similarity, *Psychological Review*, 84(4), 327–352, 1977.
14. Wang, W.J., New similarity measures on fuzzy sets and fuzzy elements, *Fuzzy Sets and Systems*, 85, 305–309, 1997.
15. Williams, J. and N. Steele, Difference, distance, and similarity as a basis for fuzzy decision support on prototypical decision classes, *Fuzzy Sets and Systems*, 121, 35–46, 2002.
16. Xie, W.X. and Bedrosian, S., An information measure for fuzzy set, *IEEE Transaction on Systems, Man, and Cybernetics*, 14(1), 1984.
17. Xuecheng, L., Entropy, distance measure and similarity measure of fuzzy sets, *Fuzzy Sets and Systems*, 52(3), 305–318, 1992.
18. Yager, R.R., On the measure of fuzziness and negation. I. Membership in unit interval, *International Journal of General Systems*, 5, 221–229, 1979.

4

Fuzzy Image Preprocessing

4.1 Introduction

Image preprocessing is the operation performed on an image at the lowest level before going for further processing which leads to an improvement of image data. The aim of preprocessing is to enhance or highlight the important features that are not properly visible and to suppress unwanted information that are not relevant to image-processing tasks. It does not increase the information content of an image. Image-preprocessing methods can be given as follows:

1. Image enhancement as a process of brightness transformation
2. Image filtering that suppresses the noise or other fluctuations that correspond to the high-frequency components in frequency domain

Enhancement algorithms are used to reduce the noise present in the image and increase the contrast of the structures in an image. For low-intensity image, where the objects of interest are not properly visible, accurate interpretation can become difficult if noise levels are high and the noise will appear as bright spots in the image. In those cases, enhancement and filtering will improve the quality of the image and provide a clear image to the human observer. The main aim is to generate a new image that is more suitable for application than the original image. We will first discuss the image enhancement in detail and then cover the filtering method.

Image enhancement for brightness transformation valid in many practical cases can be position-brightness correction and gray-scale transformation. Position-brightness correction modifies the pixel brightness by taking into account the pixel position in the image. This method assumes a linear transformation but is not valid in many practical cases because the brightness is limited to an interval scale.

Gray-scale transformation—It changes the pixel brightness but it does not take into account the position of the pixels in the image. Gray-scale transformation is simply a transformation of the gray scale to another scale to increase the contrast. The goal of the transformation is to improve the visual appearance of an image.

$$f(x,y) = S(x,y)$$

Gray-level transformations can be straight line, logarithmic, or exponential. Straight line transformation b, as shown in Figure 4.1, is a very common

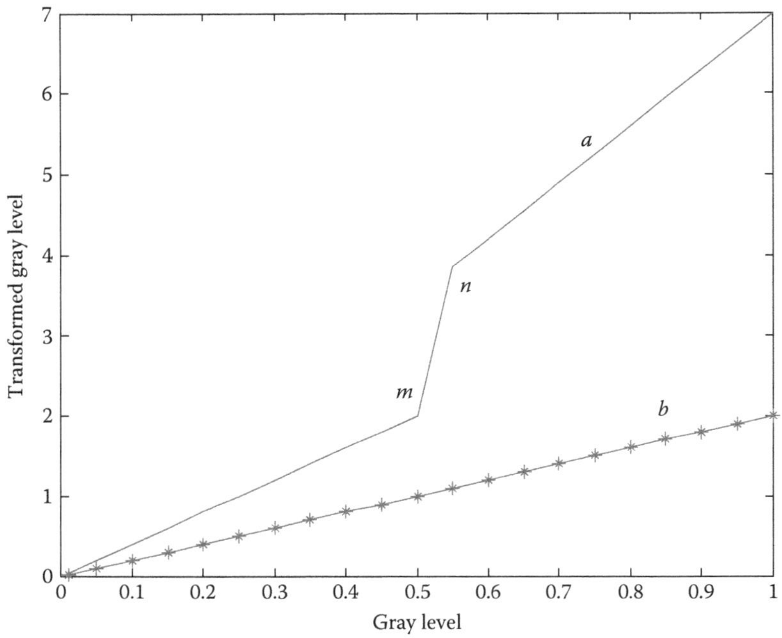

FIGURE 4.1
Gray-scale transformation.

transformation. Another common transformation is transforming one part of gray level to a certain level and another part to another level as indicated by a in Figure 4.1. The user can modify the two control points in the figure indicated by m, n to change the shape of transformation. This type of transformation is required when different gray levels require different transformations to enhance the important features of the image. It is also called contrast stretching as explained in Section 4.3.

Gray-scale transformation is histogram transformation for intensity mapping. The histogram of an image represents the frequency of occurrence of the gray levels in an image. Histogram modeling techniques are powerful enhancement techniques that modify an image to a desired shape. This is useful in stretching low-contrast image. There are several methods of histogram manipulation through intensity mapping function [18], such as the sliding, the stretching, the linear, the logarithmic, the exponential, the square root, and the equalization methods.

Histogram sliding—It involves the addition or subtraction of constant value, b, to the pixels in an image.

$$G(x,y) = f(x,y) + b$$

If $b > 0$, the brightness is increased and if $b < 0$, the brightness is decreased.

Stretching—It is the multiplication or division of all pixels by a constant value.

$$G(x,y) = a \cdot f(x,y)$$

If $a > 1$, the brightness is increased and if $a < 1$, brightness is decreased.

Logarithmic and exponential—It is a nonlinear mapping where the contrast is modified depending on the intensity value. It uses either the logarithmic function or the exponential function for histogram mapping.

4.2 Contrast Enhancement

Contrast is a property that is based on human perception. An approximate definition of contrast is

$$C = (A - B)/(A + B)$$

where A and B are the mean gray levels of the two regions where the contrast is calculated.

If the contrast between the image parts is more, the image will be more enhanced. Contrast enhancement is applied to the regions where the contrast between the regions is small but no contrast enhancement is required where the contrast is sufficient. Put differently, contrast enhancement is the gray-level transformation in such a way that the dark pixels appear darker, while the light pixels appear brighter. The aim is to generate an image of higher contrast than the original image by giving larger weights to the gray levels that are closer to the mean gray level of an image than those that are farther from the mean.

Contrast enhancement is also possible using a histogram equalization technique. It assigns the intensity values of pixels in the input image such that the output image contains a uniform intensity distribution. It increases the dynamic range of the histogram by spreading the gray values and allows us to see a large range of values. Suppose, the histogram contains many peaks and valleys, it will still have the peaks and valleys but are shifted in the resulting histogram-equalized image. This is achieved by using the normalized cumulative histogram as the gray scale mapping function.

4.3 Fuzzy Image Contrast Enhancement

Fuzzy image enhancement is based on gray-level mapping from a gray plane into a fuzzy plane using a membership transformation. It employs the principle of contrast stretching where the image gray levels are transformed in such a way that dark pixels appear much darker and bright pixels appear much brighter. The principle of contrast stretching depends on the selection of threshold, T, so that the gray levels below the threshold T are reduced and the gray levels above the threshold T are increased in a nonlinear manner. This stretching operation induces saturation at both ends (gray levels) [1].

$$\begin{aligned} \mathrm{CONT}(x) &= (1-a)\cdot x, \quad && x > t \\ &= x && x = t \\ &= (1+b)\cdot x && x > t, t \text{ is a certain threshold} \end{aligned}$$

CONT(x) is an intensifier stretching that transforms a gray level x to an intensified level. a, b are the levels between 0 and 1 that decide the percentage stretching of gray level x for a certain threshold "t." This can be extended to multiple thresholds where different regions are stretched in different ways depending on the quality of an image. The fuzziness in an image can be due to

1. Gray-level imprecision, that is, uncertainty in deciding whether a pixel is darker or brighter
2. Vague region boundary and object edges, that is, uncertainty in distinguishing the boundary between the segments or the edges of an object

Fuzzy methods for contrast enhancement employ membership values $\mu(x)$ $\mu(x)$ to know the degree of brightness or darkness of the pixels in an image. So, initially membership function is used to find the membership values of the pixels in an image that lie in the interval [0,1] using any membership function. Then a transformation is applied on the membership values to generate new membership values of the pixels in the image. Finally, an inverse transformation is applied on the new membership values for transforming the membership values back to a spatial domain. The principle of fuzzy contrast enhancement is illustrated in Figure 4.2.

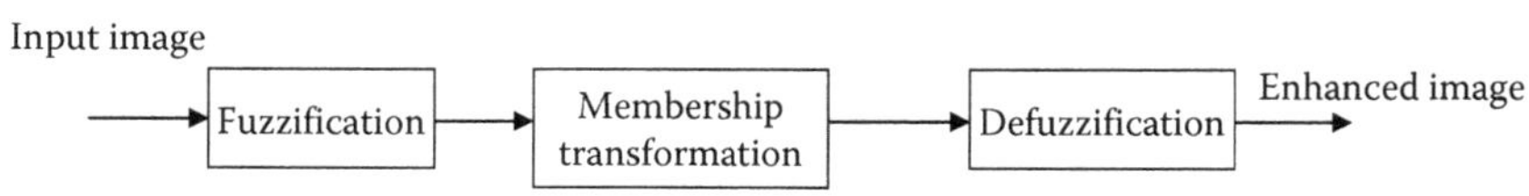

FIGURE 4.2
Fuzzy image enhancement.

Algorithmically, it can be expressed as

$$\mu'(x) = \psi(\mu(x))$$

$$x' = f^{-1}(\mu'(x))$$

where
$\mu(x)$ is the membership function
$\psi(\mu(x))$ is the transformation of $\mu(x)$ denoted as $\mu'(x)$

The purpose of contrast enhancement is to increase the overall visual appearance of the image that the human eye can visualize clearly, to be a better suited for further analysis. It is useful particularly when the intensity of any important regions of an image is such that it is difficult to make out the structures with the human eye, especially when the image is a low-contrast image, that is, medical images of tissues, blood vessels, etc. This will highlight the areas of low intensity thus improving the readability.

An image of size $M \times N$ and L gray levels can be considered fuzzy by assigning a membership function of image in terms of edginess, textural property, and darkness. In recent years, many researchers have applied various fuzzy methods for contrast enhancement. A few fuzzy methods are discussed here along with some examples.

4.3.1 Contrast Improvement Using an Intensification Operator

This method uses intensification operator to reduce the fuzziness and to increase the image contrast [8,17,18,27]. For an image I, let the gray level at location (m, n) be given by g_{mn}, and the maximum and minimum value of the gray level be given by $g_{\max}$ and $g_{\min}$. The membership function is defined as

$$\mu_{mn} = \left[1 + \frac{g_{\max} - g_{mn}}{F_{\mathrm{d}}}\right]^{-F_{\mathrm{e}}}$$

where F_{d} and F_{e} are the denominational and the exponential fuzzifiers, respectively that controls the amount of grayness ambiguity.

Numerically, as $\mu_{mn} \to 1$, when $g_{mn} \to g_{\max}$, implies that as the membership value tends to 1, that is, maximum, the gray level also approaches to maximum brightness. So it represents the extent to which the pixel possesses a bright or light gray level. The maximum membership is 1 when the gray level approaches maximum value. So $\mu_{mn} = 1$ represents maximum brightness and $\mu_{mn} = 0$ represents absolute darkness.

Membership values are then modified using intensifier operators. The intensifier operator stretches the contrast between the membership values.

It transforms the membership values that are above 0.5 (default value) to much higher values and the membership values that are lower than 0.5 to much lower values in a nonlinear manner to obtain a good contrast in an image. The stretching function is more nonlinear as the membership values move away from 0.5 to both the higher and lower values, as shown in Figure 4.3. The intensifier operation (INT) is written as

$$\mu'_{mn} = 2\cdot[\mu_{mn}]^2 \quad 0 \le \mu_{mn} \le 0.5$$
$$= 1 - 2\cdot[1-\mu_{mn}]^2 \quad 0.5 \le \mu_{mn} \le 1$$

Once the membership values are modified, modified gray values are transformed to spatial domain using an inverse function:

$$g'_{mn} = G^{-1}(\mu'_{mn})$$
$$= g_{\max} - F_d\left[(\mu'_{mn})^{-\frac{1}{F_e}} - 1\right]$$

This INT operator depends only on the membership function. It needs to be applied continuously to obtain a desired, enhanced image. This limitation is

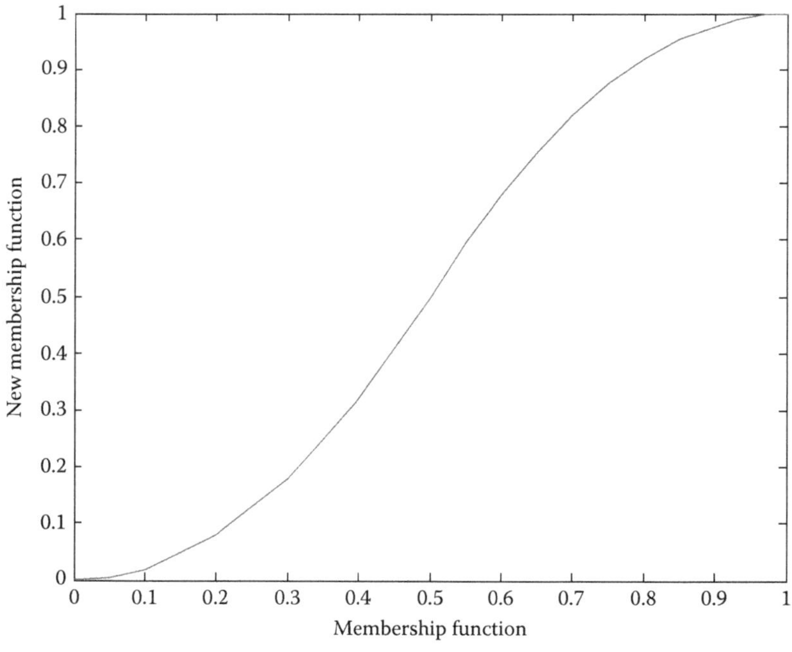

FIGURE 4.3
New membership function using INT operator.

removed using a NINT operator, reported by Handmandlu et al. [7], which uses a sigmoid function. The membership function, $\mu_X(k)$, is calculated using Gaussian membership function that uses a fuzzifier f_h i.e.,

$$\mu_X(k) = e^{-(x_{\max}-k)^2/2f_h^2}, \quad k = 0,1,2,\ldots, L-1 \text{ (gray level of the image)}$$

This membership function is then modified using a sigmoid function as follows:

$$\mu'_X(k) = \frac{1}{1+e^{-t(\mu_X(k)-0.5}}, \quad t \text{ is any parameter}, \quad k=0,1,2,\ldots,L-1.$$

The fuzzifier f_h is calculated by minimizing the fuzzy contrast and 't' is computed globally by minimizing the fuzzy entropy. Figure 4.4 is the plot of membership and modified membership function using a NINT operator at various values of t. The NINT operator does not change uniformly. For values $t \geq 5$, the NINT operator has a similar response as that of the INT operator and for values $t < 5$, the NINT operator has no appreciable influence on the membership function. Another interesting point is that if the values

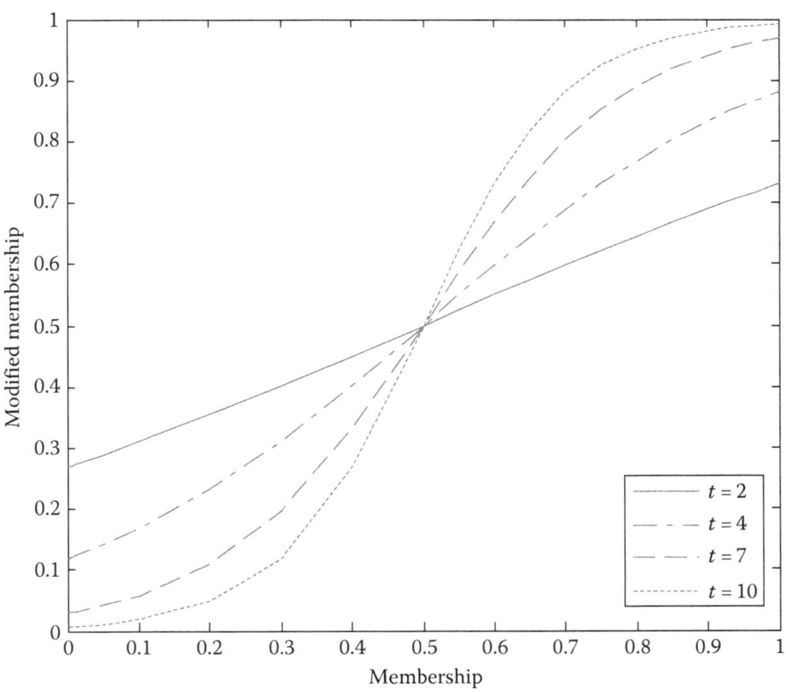

FIGURE 4.4
Membership vs. modified membership.

of t are increased by 1, the gap between the successive curves is less than that obtained using the INT operator. Thus, minute changes in the level of enhancement are possible using the NINT operator.

Example 4.1

Figure 4.5a is a CT-scanned brain image with soft tissues and ventricles. Using the low-resolution human brain image (190×190) we will illustrate the effect of INT and NINT operators on the brain image. Figure 4.5b shows the enhanced image using the INT operator. Figure 4.5c shows the result using the NINT operator.

4.3.2 Contrast Improvement Using Fuzzy Histogram Hyperbolization

The concept of histogram hyperbolization and fuzzy histogram hyperbolization is given by Tizoosh [25]. Due to the nonlinear human brightness perception, this algorithm modifies the membership values of the gray levels into logarithmic function. The procedure for histogram hyperbolization is as follows.

Initially the shape of membership function is selected based on the user's requirement. Membership values, $\mu(g_{mn})$, g_{mn}, are the gray levels of an image, and are calculated using the membership function. The fuzzifier beta, β, which is a linguistic hedge, is set in such a way that it modifies the user's membership function. Hedges [6,28,29] can be very bright, very very bright, medium bright, and so on, and the selection is made on the basis of image quality. The value of beta can be in the following range: $\beta \in [0.5, 2]$. If $\beta = 0.5$, the operation is dilation and if $\beta = 2$, the operation is concentration. If the image considered is a low-intensity image, then the fuzzifier β after operating on the membership values will produce slightly bright or quite bright images.

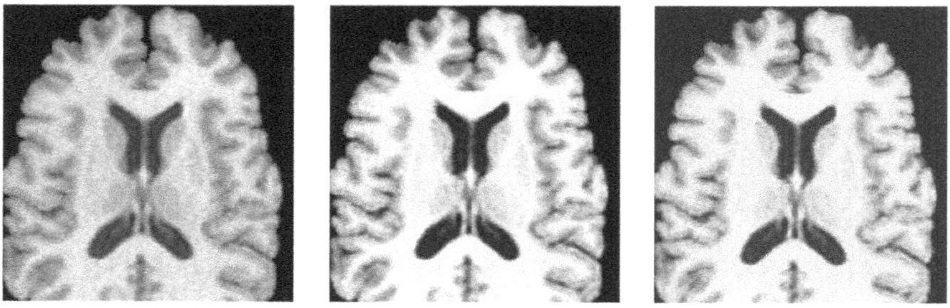
(a) (b) (c)

FIGURE 4.5
(a) Brain image, (b) enhancement using INT operator, and (c) enhancement using NINT operator.

Then new gray levels using linguistic hedges are generated using the following equation:

$$g'_{mn} = \left(\frac{L-1}{e^{-1}-1}\right) \cdot \left[e^{-\mu(g_{mn})^{\beta}} - 1\right]$$

where L is the maximum gray level of an image.

4.3.3 Contrast Enhancement Using Fuzzy IF–THEN Rules

Image quality can be improved by using human knowledge. The quality evaluation by the human observer is highly subjective in nature. Different observers judge the image in different ways. Fuzzy rule–based approach is such a method that incorporates human intuitions that are nonlinear in nature. As it is difficult to define a precise or crisp condition under which enhancement is applied, a fuzzy set theoretic approach is a good approach to this solution. A set of conditions are defined on a pixel for image enhancement and these conditions will form the antecedent part of the IF–THEN rules. The conditions are related to the pixel gray level and also to the pixel neighborhood (if required). Fuzzy rule–based systems make soft decisions on each condition, then aggregate the decisions made, and finally make decisions based on aggregation. Russo [19,20] proposed rule-based operators for image smoothing and sharpening. Choi and Krishnapuram [5] suggested three types of filters based on fuzzy rules that are used in image enhancement. Hassanein and Amr [8] used a comparative study on a mammogram image using contrast intensification method, hyperbolization and IF-THEN rules to highlight the important features of the image. A typical rule in a traditional rule-based system is

If (Condition) then (Action)

A very simple fuzzy rule–based algorithm is depicted. Initially the maximum and minimum gray levels are initialized and mid gray level is calculated. Then fuzzification of gray levels or the membership values are calculated for the dark, gray, and bright regions.

A simple inference mechanism that modifies the membership function is used as follows:

If image is dark then black

If image is gray then gray

If image is bright then white

Finally the enhanced output is computed by defuzzifying all the three regions i.e., dark, gray, and bright regions. Defuzzification is done by using the inverse of the fuzzification.

4.3.4 Contrast Improvement Using a Fuzzy Expected Value

Fuzzy expected value (FEV) by Schnieder [21] replaces the mean and median value when treating with fuzzy sets. The use of FEV in image enhancement is proposed by Schnedeir and Craig [22]. Instead of calculating the average value of a set of numbers, they evaluated a more representative value of a set. This value would indicate a typical grade of membership of a fuzzy set.

Consider a fuzzy set A in a finite set $X=\{x_1, x_2,\ldots, x_n\}$ with a membership function μ_A:[0, 1].

Let $\xi_T=\{x|\mu_A(x)\geq T\}$, $0\leq T\leq 1$, represent a subset whose elements are above or equal to the value of the threshold T. Then the fuzzy measure defined on the fuzzy subset is

$$M(\xi_T)=\frac{[\text{number of elements } x:\mu_A(x)\geq T]}{N},$$ N is the number of elements in a set.

The FEV of $\mu_A(x)$ over the fuzzy set

$$\text{FEV}=\sup_{0\leq T\leq 1}\{\min[T, M(\xi_T)]\}$$

But FEV does not generate a typical value in some cases. Schneider and Craig [23] suggested weighted FEV (WFEV), which gives a most typical value of the membership function $\mu(x)$, where weights are applied. The weight is calculated as follows.

The weighted fuzzy expected value (WFEV) is given as:

$$s=\frac{\sum_{i=0}^{L-1}\mu_i e^{(-\beta|\mu_i-s|)}.n_i^{\lambda}}{\sum_{i=0}^{L-1}e^{(-\beta|\mu_i-s|)}.n_i^{\lambda}},$$ $i=0,2,3,\ldots,L-1$ is the gray levels.

where

λ is the weight (constant) and $\beta>0$ is also a constant
n_i is the frequency of occurrence of ith gray level
μ_i is the membership value of the ith gray level

The equation for 's' is in the form $x=F(x)$ and is solved iteratively starting with $x_{n+1}=F(x)$, where x_0=FEV. This WFEV is used for image enhancement, and when applied on the gray level represents the most typical gray level. If M is a typical gray level corresponding to WFEV or FEV and g_i is the gray level, then the distance measure between M and gray level, g_i is given by

$$D_i=\sqrt{(M)^2-(g_i)^2}$$

The new gray level g_i' is computed as

$$g_i' = \max(0, M - D_i), \quad \text{for } g_i < M, \text{ when } i = 0,1,2,\ldots,M-1$$

$$g_i' = \min(L-1, M + D_i), \quad \text{for } g_i > M, \text{ when } i = M+1,\ldots,L-1$$

$$g_i' = M, \quad \text{otherwise}$$

g' is the final contrast-enhanced image.

Example 4.2

As in Example 4.1, we will show an example of a low-resolution human brain image in Figure 4.6 that will illustrate the effect of histogram hyperbolization, fuzzy IF-THEN rules, fuzzy expected value on the brain image.

4.3.5 Locally Adaptive Contrast Enhancement

Apart from global enhancement, locally adaptive enhancement algorithm selects an $m \times n$ neighborhood of a pixel and moves the center of this area

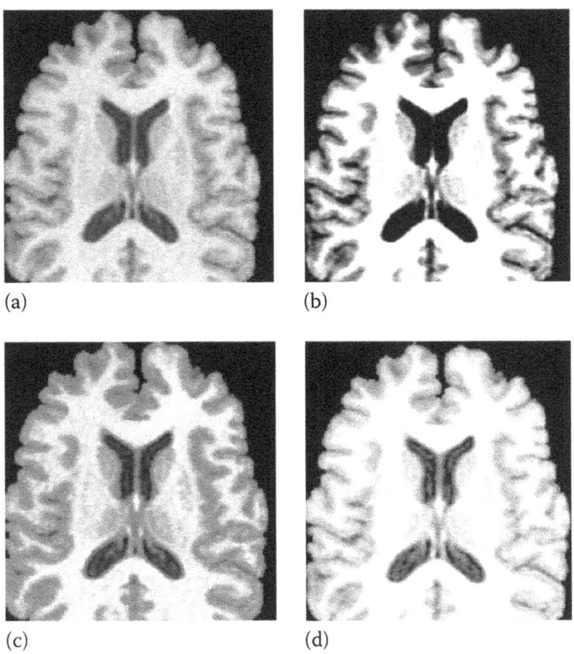

FIGURE 4.6
(a) Brain image, (b) enhancement using histogram hyperbolozation, enhancement using IF-THEN rules, and (c) enhancement using FEV.

(window) from pixel to pixel and at each location enhancement the algorithm is applied. Local contrast enhancement increases the performance of enhancement procedure. As the images have variable and different gray levels, global histogram treats all regions of an image equally and thus yields poor local contrast enhancement. For medical images, especially, where the image is of poor contrast and low quality, locally adaptive enhancement works well. In locally adaptive enhancement method as described, the image is divided into subblocks, and enhancement algorithms are applied for each subblock. In this method, the minimum and maximum gray levels are computed that may lead to noise, thereby affecting the membership values. This can be improved by increasing the number of subblocks.

4.4 Filters

When the images are transmitted over channels, they are corrupted with impulse noise due to noisy channels. This impulse noise consists of large positive and negative spikes. The positive spikes have values much larger than the background and thus they appear as bright spots, while the negative spikes have values smaller than the background and they appear as darker spots. Both the spots for the positive and negative spikes are visible to the human eye. Also, Gaussian type of noise affects the image. Thus, filtering for both impulse noise and Gaussian noise is required before processing. There are lots of classical and fuzzy filters in the literature to remove noise. The classical filters are the mean filter, the weighted filter, the adaptive weighted filter, the adaptive Wiener filter, the Gaussian filter, the median filter, the adaptive median filter, and the extended mean filter. A brief overview of classical filters is provided in the following text.

The mean filter or the averaging filter helps in smoothing operations. It suppresses the noises that are smaller in size or any other small fluctuations in the image. It involves in calculating the average brightness values in some neighborhood $m \times n$ and replaces the gray level with an average value. Smoothing or averaging operation blurs the image and does not preserve the edges. These are not used in removing noise spikes. For a 3×3 neighborhood, the convolution mask is used as

$$\frac{1}{9}\begin{bmatrix} 1 & 1 & 1 \\ 1 & 1 & 1 \\ 1 & 1 & 1 \end{bmatrix}$$

Adaptive weighted mean filter [16] is similar to mean filter where the gray level is replaced by a weighted average of the gray values. Weights are calculated from the gray-level difference. If this difference exceeds a certain threshold, then the pixel is a noise pixel.

Adaptive Weiner filter [16] replaces the center value of the pixel by a sum of the local mean value and a fraction of contrast, where this fraction depends on the local estimation of the variance.

Gaussian filter can be used for smoothing operations. A 2-D Gaussian smoothing operator $G(x, y)$ is given by

$$G(x,y) = \exp\left\{-(x^2 + y^2)/2\sigma^2\right\}$$

where x and y are the image coordinates and σ is the standard deviation. The standard deviation, σ, is proportional to the size of the neighborhood. Gaussian filter has some characteristics. The pixels that are nearer, that is, within a distance, σ, have a good influence, and if farther than 3σ, the pixels have negligible influence. This provides better suppression of noise than rectangular or triangular filter.

Median filter is a nonlinear smoothing method. It calculates the median of the some $m \times n$ neighborhood and replaces the pixel with this median value. This is done for the whole image. It does not blur the image and eliminates the noise spikes or the impulse noise. The filter can be applied iteratively to obtain better quality image. But when the impulses are high, median filter cannot work well and in that case adaptive median filter that uses variable window sizes can be useful. Median filters do not smooth the image.

An adaptive nonlinear filter is more efficient than a median filter when the impulse noise is high. It uses variable window sizes for removing mixed impulses, that is, negative and positive impulses, and restores sharpness. Hwang and Haddad [9] suggested two adaptive median filters: rank order–based adaptive median filter (RAMP) and impulse size–based adaptive median filter (SAMP).

RAMP is based on the test for presence of noise at the center pixel followed by a test for the presence of any residual impulse at the median filter output. It has two levels in the first level, median filter is applied on the image level and tests the output of the filter if any residual impulse is present. If no impulses are present in the filter output, then a second level test is performed. The second level test checks whether the pixel itself is corrupted with impulse. If it is not corrupted with impulse noise, then it is left without further filtering. If in the first level test, there is impulse in the median filter output, then window size is increased and the first level test is repeated.

SAMP is based on the size of impulse noise that adjusts the size of the window. It consists of two operations, that is, detection and filtering. At each location, impulse noise is tested and then removed with a median filter. This is a three-stage algorithm.

4.5 Fuzzy Filters

Fuzzy filters provide promising results in image-processing tasks that cope with some drawbacks of classical filters. Fuzzy filter is capable of dealing with vague and uncertain information. Sometimes, it is required to recover a heavily noise corrupted images where a lot of uncertainties are present and in this case, fuzzy set theory is very useful. Each pixel in the image is represented by a membership function and different types of fuzzy rules that considers the neighborhood information or other information to eliminate noise, are used. Fuzzy filters are very robust in the sense that the classical filter removes the noise with blurry edges but fuzzy filters perform both the edge preservation and smoothing.

Images and fuzzy set can be modeled in a similar way. Fuzzy set in a universe of X is a mapping from X to a unit interval [0,1], that is, every element x of X is associated with a membership degree. Similarly, in the normalized image where the image pixels ranging from {0, 1, 2,…, 255} are normalized by 255, the values obtained are in the interval [0, 1]. Thus, it is a mapping from the image G to [0, 1]. In this way, the image is considered as a fuzzy set and thus fuzzy filters are designed.

Noise reduction is an important area for image preprocessing. Besides classical filters, there are lots of fuzzy filters in the literature. Images can be corrupted with impulse noise, Gaussian noise, or both. Depending on the type of noise, filters can be used. The fuzzy filters are categorized into two subclasses: (a) fuzzy-classical filters and (b) fuzzy filters. Here we discuss the various types of fuzzy filters briefly.

1. *Fuzzy-classical filters* are filters that use fuzzy logic and these are the modifications of the classical filters. Some fuzzy-classical filters are mentioned below.
   a. Fuzzy median filter—Fuzzy median filter [3,11] is well known for removing impulse noise. It is the fuzzy rank ordering of samples and is simply a replacement of conventional median filter with fuzzy counterparts.
   b. Fuzzy impulse noise detection and reduction method—This filter by Schulte [24] detects the impulse noise and any other noise in the image. It contains the noise-detection step and filtering step to preserve the edges. Fuzzy detection step uses fuzzy gradient values in eight directions within a 3×3 window, which indicates the degree of central pixel as an impulse noise pixel. A fuzzy set is constructed based on the gradient values which is represented by a membership function such as BIG, SMALL, BIG POSITIVE, and BIG NEGATIVE. Then a fuzzy rule is applied to decide if the control pixel as a noise pixel.

c. Fuzzy weighted mean filter—This filter is an extension of the adaptive weighted mean filter. The weights take the values in the interval [0,1] instead of the values 0 and 1. The weights are determined by using fuzzy rules, which are based on the difference or the distance between a pixel and its neighborhood.

d. Weighted fuzzy mean filter (WFM)—In this filter [12,13,16], weights takes the values in the form of membership function. The detection of noise pixels is based on the observation of the noise pixel in the homogeneous region. If the pixels in a region are dark, then the noise pixel will not be dark and consequently the membership degree corresponding to this noise pixel is small. These membership degrees are used as weights. This is done for all the bright, less bright, and dark pixels. Three fuzzy weighted means are calculated in the fuzzy sets of dark, medium, and bright, and a triangular membership function is used. Then the decision process calculates which of the three fuzzy weighted means will be the output.

   First adaptive weighted fuzzy mean filter (AWFM) is an extension of the WFM filter where the membership functions are determined adaptively instead of fixed membership function.

   Second adaptive weighted fuzzy mean filter is also an extension of the WFM. The drawback in the WFM is that it removes heavy additive impulse noise, but when the noise probability is lower than 0.1, the WFM does not work well. Thus the AWFM consists of two more operators: (a) fuzzy detectors and (b) dynamic selectors.

e. Choi and Krishnapuram filter—This type of filter [4] removes impulse noise, smoothes out non-impulse noise, and enhances edges or certain other salient features. Three different filters are designed for each task using the weighted least mean squared method. The criteria for selecting each filter are based on the local context and on particular applications. Few fuzzy rules are constructed based on the criteria and the corresponding filters constitute the consequents of the fuzzy rules. The overall result of the fuzzy rule–based system is computed by combining the results of each filter.

f. Fuzzy decision–directed filter—Mancuso et al. [14] proposed this filter to detect the impulsive noise and is performed only when noise is detected. This approach considers the edge and motion information to preserve relevant structures. A fuzzy process is developed that detects the noise configuration and a classical filter is activated depending on the output of the structure-extraction process. The idea is to make the impact of the classical filter dependant on the degree to which a pixel is considered as a noise pixel.

g. Fuzzy multilevel median filter—This type of filter [2,31] preserves the basic structures of the image. A window of size $(2N+1)\times(2N+1)$ centered at (i, j)th is selected and four subsets of the square window at (i, j)th point is defined. These windows show one dimensional line segment. Features may span two or more line segments and the subclass may be extended to complex features. But in that, noise suppression is diminished. The median values of the four subsets are calculated and the output of the filter is designed based on the combination of the median value of four sub-windows.

h. Multipass fuzzy filter—This type of filter by Kam et al. [10] has given the idea of multipass filter that removes both impulse and Gaussian noises in images. The image is fuzzified using sigmoid membership function that considers a linguistic term between the central pixel and its neighboring pixels. Three types of membership functions are defined out of which two are for impulse noise detection and one for Gaussian noise detection. In the first pass, the noise image is passed through an impulse noise filter and in the second pass, the filtered image in the first pass is passed through a Gaussian noise filter.

Moving average filter (MAV), Gaussian moving average filter (GMAV), Gaussian fuzzy filter with median value (GMED), symmetrical triangular fuzzy filter with median centre (TMED) are the fuzzy–classical filters by Kwan [11]. For all these filters, the output of the fuzzy filter for the input $x(i, j)$ is defined as:

$$y(i,j)=\frac{\sum_{m,n\in A}F[x(i+m,j+n)]\cdot x(i+m,j+n)}{\sum_{m,n\in A}F[x(i+m,j+n)]},\quad F[(x(i,j)]$$

is the window function of size $w\times w$, A is the area of the window.

i. MAV—It is equivalent to a fuzzy filter with rectangular window of size $w_1\times w_2$ covering all input values [11].

The window function is defined as:

$$F[x(i+m,j+n)]=1,\ \text{for}\ m,\ n\in A$$

j. GMAV—Gaussian moving average value within the window at the center is calculated using Gaussian function:

$$F[x(i+m, j+n)] = \exp\left(-\frac{[(x(i+m, j+n) - x_{\text{mav}}(i, j)]^2}{2\sigma(i, j)^2}\right)$$

where $\sigma(i, j)$ and $x_{\text{mav}}(i, j)$ are, respectively, the variance and the moving average value at (i, j)th point computed from the input variables in the window.

k. GMED—The filter value within a window at the center is calculated using Gaussian function:

$$F[x(i+m, j+n)] = \exp\left(-\frac{[(x(i+m, j+n) - x_{\text{med}}(i, j)]^2}{2\sigma(i, j)^2}\right)$$

where $\sigma(i, j)$ and $x_{\text{med}}(i, j)$ are the variance and the median value at (i, j)th point, respectively, computed from the input variables in the window [11].

l. TMED—This is also similar to the GMED but the center value is defined in a different way from the Gaussian function [11].

2. *Fuzzy filters* are filters that are totally dependent on fuzzy logic and they do not have any connection with classical filters. A few fuzzy filters are discussed briefly in the following text:

   a. Gaussian noise reduction filter (GOA)—This filter is specially designed to remove Gaussian noise. Averaging is done for a pixel using other neighborhood pixels and simultaneously taking care of the other image structures such as edges. To achieve this, two features are required. First, in order to distinguish between the variations due to noise and the image structures, the filter uses gradient for all the eight directions. Second, the membership functions are adapted accordingly to the noise level to perform fuzzy smoothing. The filter is applied iteratively [15].

   b. Histogram adaptive filter (HAF)—This type of filter [30] removes high impulsive noise, preserving edge information. In HAF, each input pixel is considered a fuzzy variable and a square window of size 3×3 is slided over the entire image and the filter output is associated with each center pixel in a window. Three fuzzy sets for dark, medium, and bright are created and the membership functions for these fuzzy sets are calculated. Then fuzzy inference rules based on the Takagi–Sugeno approach with a slight difference is used in a final output decision process.

c. Iterative fuzzy control–based filter (IFCF)—This filter is used to remove both impulse and Gaussian noise. The idea behind the filter is that if the gray value differences between the processed pixel and its neighbors belong to a certain class is more, then the correction term should belong to that class. For the construction of IFCF, seven classes of different gray value differences are considered. For a triangular fuzzy set, the classes are negative big, negative medium, negative small, zero, positive small, positive medium, and positive big. Then the gray-level differences of the processed pixel and its neighbors are computed and the fuzzy rules are applied. In this filter, the image is somewhat blurred [15].

   Modified iterative fuzzy control–based filter (MIFCBF) and extended iterative fuzzy control–based filter (EIFCBF) are modified versions of IFCFs and these are designed to counter the blurriness in the image.

d. Smoothing fuzzy control–based filter (SFCF)—This filter is a noniterative version of the IFCF and is implemented with two modifications. The modifications are in the fuzzy rule and in the activation degree ('average', is replaced by 'median') in IFCF. Sharpening and smoothing fuzzy control filter (SSFC) is an extension of SFCF where the sharpening of edges is considered. For this, two more fuzzy rules are added to make the edges more sharp [15].

e. Fuzzy inference rule by else-action filter (FIRE)—These are nonlinear operators by Russo and Ramponi [19,20] that process image data using fuzzy rules to calculate the degree or the correction term whether the pixel is a noise pixel. For each pixel in a noisy image, neighboring pixels in the window of size 3×3 are considered and the FIRE operator processes the neighborhood information by using fuzzy rules to estimate the correction term. Fuzzy triangular membership function is used to find the membership values of the pixels. If none of the rules are satisfied, then the central pixel is left unchanged. The output variable is a correction term that cancels the noise. FIRE filters are applied recursively to increase the performance i.e., the processed pixel is replaced immediately by the filtered pixel.

   PWLFIRE: Piecewise linear fuzzy inference ruled by else-action filter [20] follows the same concept as FIRE filters but with two modifications. First, the filter uses two linear piecewise fuzzy sets instead of linear or triangular fuzzy sets (as in FIRE), and these sets depend on a several parameters. Second, neutralization of small corrections is performed by incorporating the choice of the membership function parameters.

f. Fuzzy similarity–based filter (FSB)—This filter by Tolt and Kaylakov [26] is based on fuzzy similarity and homogeneity. When an image is corrupted with noise, many pixels change their values from their original values. In such cases, the noise cancellation task is carried out by analyzing the pixel similarity in all the local neighborhoods by scanning the whole image. Each local neighborhood is represented by a window of size $a \times b$. As the true pixel value is unknown, a set of templates corresponding to different local image properties are used to analyze the intensity distribution in the window. Each template is formed by the central pixel x_0 and the number of window pixels x_i. Several fuzzy rules are then applied that compare the central pixel, x_0 with the templates using a fuzzy relation "similar."

g. Fuzzy cluster filter (FCF)—This is a fuzzy weighted mean filter applied iteratively. Weights are dependent on the distance to the cluster center.

4.6 Summary

This chapter describes image preprocessing that is required before further processing. Preprocessing is done to enhance or highlight the image structures and suppress unwanted information in the image. Image enhancement and filtering are described in this chapter. Enhancement schemes are used to make the contrast of the image higher and clearer. Also various types of fuzzy enhancement schemes proposed by different authors are discussed. Also, various types of filters are discussed to remove the noise in the image. Filters may be crisp or fuzzy. In this chapter, fuzzy filters are described. Fuzzy-classical filters, which are the modifications of classical filters using fuzzy logic, and fuzzy filters, which use only fuzzy logic and are not in connection with classical filters, are briefly discussed.

References

1. Acharya, T. and Ray, A.K., *Image Processing: Principles and Application*, Wiley, Hoboken, NJ, 2005.
2. Arce, G.R. and Foster, A.R., Multilevel median filter: properties and efficacy, in *Proceedings of the ICASS88*, vol. 2, New York, pp. 824–827, 1988.
3. Bas, J. and Neira, A.P., Fuzzy median filter for CDMA detection for non Gaussian channels, in *IEEE Proceedings in Vehicular Technology Conference*, vol. 2, Atlantic City, NJ, pp. 1210–1214, 2001.

4. Choi, Y. and Krishnapuram, R., Image enhancement based on fuzzy logic, in *Proceedings of the International Conference on Image Processing*, vol. 1, pp. 167–171, 1995.
5. Choi, Y. and Krishnapuram, R., A robust approach to image enhancement on fuzzy logic, *IEEE Transaction on Image Processing*, 6(6), 808–825, 1997.
6. Gonzales, R.C. and Woods, R.E., *Digital Image Processing*, Addison-Wesley, Reading, MA, 1992.
7. Handmandlu, M. et al., Color image enhancement using fuzzy intensification, *Pattern Recognition Letters*, 24, 81–87, 2003.
8. Hassanien, A.E. and Amr, B., A comparative study on digital mammography enhancement algorithm based on fuzzy set theory, *Studies in Information and Control*, 12(1), 21–31, 2003.
9. Hwang, H. and Haddad, R.A., Adaptive median filters, New algorithms and results, *IEEE Transactions on Image Processing*, 4(4), 499–502, 1995.
10. Kam, H.S. et al., A new tunable fuzzy filter for image smoothing, in *Proceedings of the International Symposium on ISSPIT 2003*, Darmstadt, Germany, pp. 680–683, 2003.
11. Kwan, H.K., Fuzzy filters for noisy image filtering, *Proceedings of the IEEE*, 4, 161–164, 2003.
12. Lee, C.S. and Kuo, Y.H., Adaptive fuzzy mean filter and its application to image enhancement, in *Fuzzy Techniques in Image Processing*, Kerre, E.E. and Nachtagael, M. (Eds.), Springer-Verlag, New York, 2000.
13. Lee, C.S., Kuo, Y.H., and Yu, P.T., Weighted fuzzy mean filters for image processing, *Fuzzy Sets and Systems*, 89, 1997, 157–180.
14. Mancuso, M. et al., A fuzzy decision directed filter for impulsive noise reduction, *Fuzzy Sets and Systems*, 77, 111–117, 1996.
15. Nachtagael, M. et al., An overview of fuzzy filters for noise reduction, in *Proceedings of the FUZZ IEEE*, pp. 7–10, 2001.
16. Nachtagael, M. et al., An overview of fuzzy and classical filters for noise reduction, in *Proceedings of the FUZZ IEEE*, pp. 3–6, 2001.
17. Pal, S.K. and King, R.A., Image enhancement using smoothing with fuzzy sets, *IEEE Transactions on Systems, Man, and Cybernetics*, SMC-11(7), 494–501, 1981.
18. Pal, S.K. and King, R.A., On edge detection of x-ray images using fuzzy sets, *IEEE Transactions on Pattern Analysis and Machine Intelligence*, PAMI-5(1), 69–77, 1983.
19. Russo, F. and Ramponi, G., Combined FIRE filters for image enhancement, in *Proceedings of the Third International IEEE Conference on Fuzzy Systems*, Orlando, FL, pp. 264–267, 1994.
20. Russo, F., Fire operators in image processing, *Fuzzy Sets and Systems*, 103, 265–275, 1999.
21. Schneider, M. and Kandel, A., Properties of fuzzy expected value and fuzzy expected interval in fuzzy environment, *Fuzzy Sets and Systems*, 28, 1988.
22. Schneider, M. and Craig, M., On the use of fuzzy sets in histogram equalization, *Fuzzy Sets and Systems*, 45, 271–278, 1992.
23. Schneider, M. and Kandel, A., The use of weighted fuzzy expected value (WFEV) in fuzzy expert systems, *Fuzzy Sets and Systems*, 31, 37–45, 1989.
24. Schulte, S. et al., A fuzzy impulse noise detection and reduction method, *IEEE Transaction in Image Processing*, 15(5), 1153–1162, 2006.

25. Tizhoosh, H.R. and Fochem, M., Fuzzy histogram hyperbolization for image enhancement, in *Proceedings of EUFIT 95*, vol.3, Aachen, 1995.
26. Tolt, G. and Kaylakov, I., Fuzzy similarity based noise cancellation for real time image processing, in *Proceedings of the 10th FUZZ IEEE*, Melbourne, Victoria, Australia, pp. 15–18, 2001.
27. Zadeh, L.A., *Fuzzy Logic and Its Application*, Academic Press, New York, 1965.
28. Zadeh, L.A., A fuzzy set-theoretic interpretation of linguistic hedges, *Journal of Cybernetics*, 2, 4–34, 1972.
29. Sonka, M., *Image Processing Analysis and Computing Vision*, Brooks/Cole, Florence, KY, 2001.
30. Wang, J.H. and Chiu, H.C., HAF: An adaptive fuzzy filter for restoring highly corrupted images by histogram estimation, *Proceedings of the National Science Council*, 23(5), 630–643, 1999.
31. Yang, X.H. and Toh, P.S., Adaptive multilevel median filter, *IEEE Transactions on Image Processing*, 4(5), 680–682, 1995.

5

Thresholding Detection in Fuzzy Images

5.1 Introduction

Thresholding is a popular, fast, and computationally inexpensive segmentation technique. If the objects in an image are disjoint and their gray levels are clearly distinct from the background, thresholding is the appropriate type of segmentation. In a text, the characters are generally darker than the paper. In pictures of mitotic cells, chromosomes or tracks are darker or lighter than the background. In all these cases, the gray-level histogram of the picture displays the peaks corresponding to the two gray levels: characters and the paper, chromosomes or tracks and background. In thresholding, an appropriate threshold is chosen that separates these peaks and segments the picture into two or more segments: one (or more) for the object(s) and the other for the background. A histogram containing a single peak is called unimodal, two peaks is called bimodal, and multiple peaks is called multimodal. A complete segmentation of an image R is a finite set of regions R_1, R_2, R_3, …, R_N such that

$$R = \bigcup_{i=1}^{N} R_i \quad \text{and} \quad R_i \cap R_j = \phi \quad i \neq j$$

Thresholding is a transformation of an input image A into a segmented output image B as follows:

$b_{ij} = 1$, for $a_{ij} \geq T$

$= 0$ for $a_{ij} \leq T$, where T is the threshold

$b_{ij} = 1$, for the image pixels that belong to the object class

$b_{ij} = 0$, for the image pixels that belong to the background class

However, the selection of the threshold is very crucial in segmenting an image. If the threshold is not selected properly, proper segmentation cannot be obtained. One such segmentation performed using threshold values too high and too low is shown in Figure 5.1. If the threshold value is more than the actual threshold, the image is oversegmented and if the threshold value is less than the actual threshold, the image is less segmented.

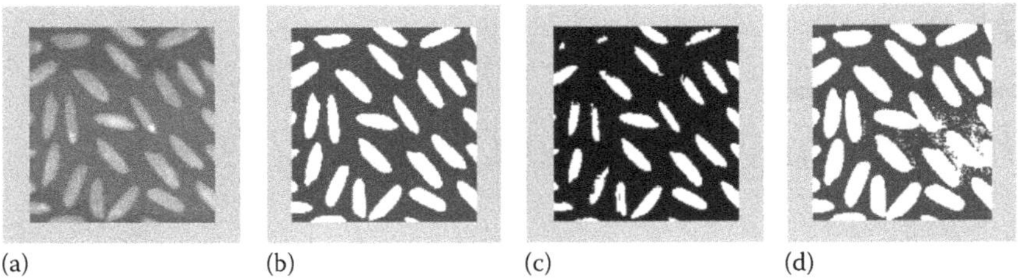

FIGURE 5.1
Image thresholding: (a) original image, (b) threshold segmentation, (c) threshold too high, and (d) threshold too low.

As described in previous chapters, in many practical situations the image pixels are not precise and the boundaries between the regions, the edges demarcating a segment, and the definition of the regions are imprecise or vague. Fuzzy thresholding techniques are more appropriate in dealing with such images and the resulting thresholded images are more accurate. Each pixel is assigned a membership degree using an appropriate membership function as mentioned in Chapter 1 and the threshold is selected by an optimum threshold selection method.

5.2 Threshold Detection Methods

There are various ways of choosing the threshold of an image. We will first mention some of the common techniques.

1. If the area covered by the objects and the background of an image are known a priori, then an easy way to threshold an image is by the P-tile method. In this method, the threshold T is selected by looking at the image histogram. In an image with a bimodal histogram, if the object area covers $(1/a)$th part of the image, threshold T is selected such that a $1/a$ portion of the image has gray values below the threshold T and the rest of the gray values are larger than the threshold T. In reality, however, one does not have such prior information about the area ratios and in that case the P-tile method is not useful.
2. There are other methods of threshold detection that are based on histogram shape analysis. Thresholding usually involves analyzing the histogram. If an image consists of distinct objects of almost homogeneous gray levels for the object and also for the background, the histogram is bimodal. In a bimodal histogram, there exists a valley between two peaks: one for the object pixels and other for the

background pixels, and the threshold value lies in the deep valley. In the case of a multimodal histogram, where the image has multiple objects of varying gray levels or intensities, there will be more peaks and more valleys. In such cases, there are multiple thresholds and each threshold lies on or near the valley between the consecutive two maxima or peaks this technique is called the mode method.

3. Quite often, the image histogram consists of two local maxima belonging to the same global maximum. In such images, one may perform histogram smoothing to detect the minima between the two maxima.
4. For better threshold determination, that is, for obtaining distinct peaks and valleys, one can weight the histogram by suppressing the pixels with a high gradient. In such cases, the histogram will contain the gray levels that belong mostly to the object and the background and may not contain the border pixels having a high pixel gradient.
5. Another technique of thresholding involves constructing a gray-level histogram that consists mostly of the border or edge pixels, that is, high gradient pixels. The histogram in such cases is unimodal and the peak corresponds to the gray levels bordering between the object and the background. The threshold is selected as the gray level of the peak.

5.3 Types of Thresholding

There are different strategies of thresholding, namely, (a) global thresholding, (b) local thresholding, (c) iterative thresholding, (d) multispectral thresholding, and (e) optimal thresholding. These strategies are discussed briefly.

5.3.1 Global Thresholding

In a global thresholding, the image is considered as a whole and the threshold, once selected, remains constant throughout the image. Depending on the modality of the histogram, the threshold levels may be single, double, or multiple.

A single threshold determines the value by treating each pixel independent of its neighborhood. In this, the histogram contains two peaks: object and background.

For a double threshold, when an image consists of two objects of different gray levels, two different threshold values are used to threshold the image. The histogram that is trimodal contains three peaks: two for the object and

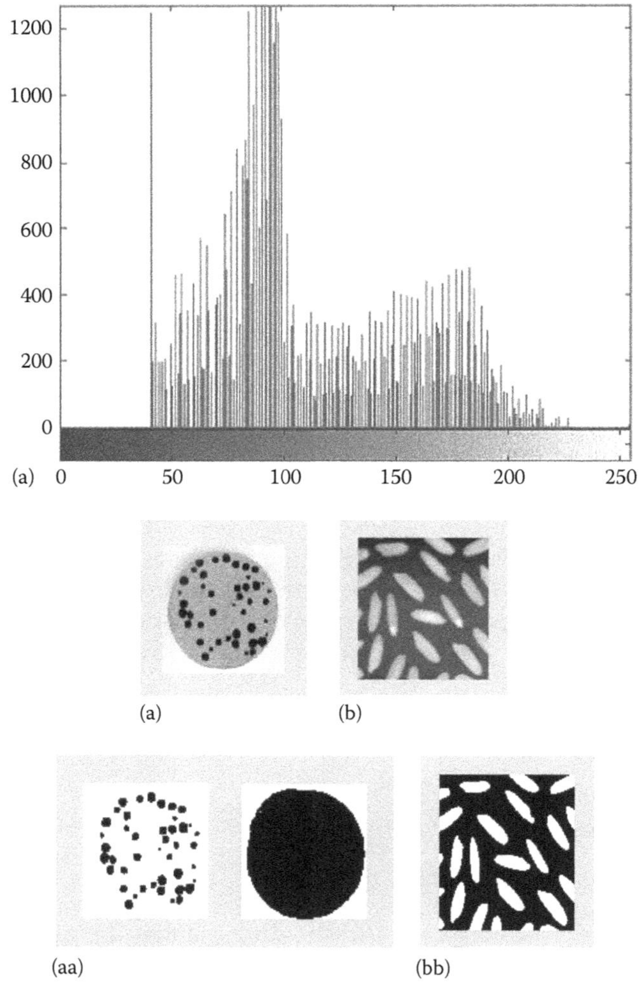

FIGURE 5.2
Global thresholding.

one for the background. An example is shown in Figure 5.2 for a global threshold that uses single and double threshold values.

5.3.2 Locally Adaptive Thresholding

In local thresholding, the image is divided into several windows or regions and thus the threshold is not constant throughout the image. Rather, the threshold values vary throughout the image and there is one threshold (or one set of thresholds) for each window. The threshold value depends on the local statistics of the region such as variance, mean, etc., of the image. It may be termed as regional thresholding. It works well when the intensities are not uniform and multiple objects are present with different gray levels. Such types of images are the real images, for example, medical images where the images are not properly illuminated. One possible way by Niblack [17] is replacing each pixel by mean and standard deviation of its local neighborhood of size $b \times b$. Threshold may be calculated as

$$T(i,j) = m(i,j) + k \cdot \sigma(i,j)$$

where m, σ, and k are the local mean, variance, and bias setting, respectively.

Another easy possible way is to divide the image into several windows and then calculate the threshold is for each window. Then the image is thresholded using all the threshold values of the windows.

5.3.3 Iterative Thresholding

Iterative thresholding is one of the iterative schemes by Leung [16] and Ridler and Calvard [25]. Initially a threshold value, T, is selected for an image using the average of the background and foreground regions. With this threshold value, the image is thresholded into two regions, R_1 and R_2. Again the mean is calculated for the two regions and a new threshold, $T_{n+1} = m_{\mathrm{T}n(\mathrm{obj})} + m_{\mathrm{T}n(\mathrm{back})}/2$, is selected, where $m_{\mathrm{T}n(\mathrm{obj})}$ and $m_{\mathrm{T}n(\mathrm{back})}$ are the means of the object and background regions at the nth iteration. The same step is repeated till there is no appreciable change in the threshold, $|T_n - T_{n-1}|$.

5.3.4 Optimal Thresholding

In an image, histogram peaks are not always clear. An image may consist of slight variations of gray levels and so prominent peaks may not be present. Peaks may overlap, and choosing a threshold in the valley between two overlapping peaks will not classify the pixels into different regions. There are various methods for threshold selection that deal with the probability densities of the image histogram using normal distributions. The gray-level distribution is modeled as a mixture of two Gaussian or normal distributions for the object and background. The threshold is selected corresponding to the minimum probability value between the maxima of two normal distributions, which results in minimum error segmentation [9]. But there is a difficulty in estimating the normal distribution parameters together with an uncertainty in considering the distribution normal. These difficulties can be eliminated by using the technique called optimal thresholding, where an optimal threshold is selected. In optimal thresholding, a criterion function is designed that yields some measure of separation between regions. The criterion function is calculated for each gray level, and the gray value that maximizes or minimizes the measure is chosen as the threshold. There are many different methods, such as maximizing the gray-level variance [19],

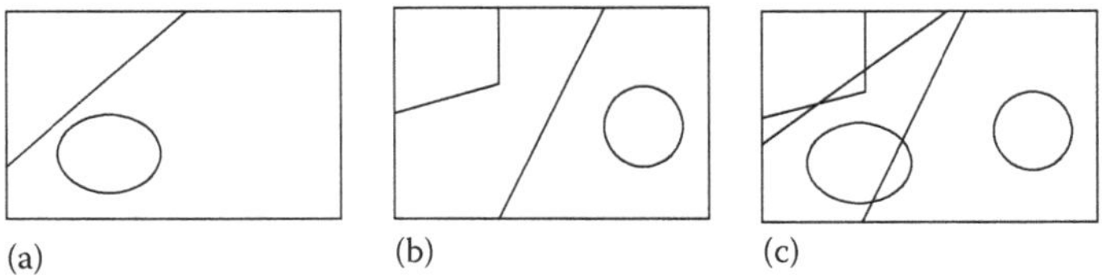

FIGURE 5.3
Multispectral thresholding: (a) band 1 thresholding, (b) band 2 thresholding, and (c) multispectral segmentation.

minimizing the classification of the error probability based on the condition that the histograms are Gaussian [15], entropy [2–4,10,15,18,19,20,26–29,34], [2,4,5,12,17,19,21,22,28–31,36], spatial entropy [1,2], similarity [6,10], and moment [33], measures of fuzziness [6,7,10,21,23].

5.3.5 Multispectral Thresholding

Many images such as color images contain three spectral (color) bands, for example, red, green, and blue. Multispectral remote sensing images or meteorological satellite images may have even more spectral bands that require multispectral thresholding. Segmentation in this case is based on n-dimensional vectors of gray levels in n spectral bands for each pixel or small pixel neighborhood. This segmentation is widely used in remote sensing. Image segmentation in this case may be achieved by using a multilevel thresholding. The simplest way to threshold an image is to obtain a threshold independently in each spectral band and combine them into a single segmented image. The algorithm is as follows [32]:

Algorithm 5.1: Multispectral thresholding

Step 1. Initialize the whole image as a single region.
Step 2. For each spectral band, compute a smoothed histogram. Find the most significant peaks in each histogram of the spectral bands and determine two thresholds as the local minima on either side of this most significant peak or the maxima. With these thresholds, segment the region in each spectral band. Each segmentation in each spectral band is then combined to a multispectral segmentation as shown in Figure 5.3.
Step 3. Repeat step 2 for each region in an image until the histogram of each region contains a single significant peak.

5.4 Thresholding Methods

We are now going to discuss different methods of image thresholding. There are many different approaches for image thresholding. Some of the approaches are related to crisp logic as well as fuzzy logic. As we are dealing with fuzzy set theory, approaches related to crisp logic are discussed in brief [28,30] and approaches related to fuzzy set are discussed in details with examples.

Histogram shape–based methods: In this method peaks, valleys, and curvatures of the smoothed histogram are analyzed. A convex hull of the histogram is constructed, which is the smallest convex polygon containing the histogram. Then the concavities are found by taking the difference between the convex hull and the histogram. The local maxima of the concavities corresponding to the position valleys of the histogram are used as threshold values [27]. But this method is very sensitive to noise as the concavities are noise sensitive. Instead of convex hull, an exponential hull method, which is a variation of convex hull for concativity analysis may be used [35]. The exponential hull is an exponential function of the upper convex hull of the histogram's logarithm. This method is less sensitive to noise. Ramesh et al. [24] used functional approximation to the probability density function. It is approximated by a two-step function where either the sum of the squares or the variance of the functional approximation is minimized.

Clustering-based methods: In this method, gray levels are clustered in two parts as background and object, which means that within a cluster the pixel gray levels are similar. So there is a similarity of gray levels within the cluster and dissimilarity of gray levels in different cluster. To perform the class separatibility, various methods are suggested by the researchers and then a criterion function is formulated to select an optimum threshold. Otsu's [19] method is based on maximizing the between class variance or minimizing the within class variance of the object and background. Otsu's method gives good result when the number of pixels in each class is close to each other. Kittler and Illingworth [14] used minimum error criterion where the minimum error threshold is derived by minimizing the probability of classification error based on the condition that a histogram follows the Gaussian distribution.

Entropy-based methods: It uses the entropy of the object-background regions, the cross entropy between the original and the binarized image, relative entropy, etc. Entropy is a measure of information contained in an image and thus serves as a measure of separation. It finds out separately the information of two regions—above and below the threshold and calculates the entropies for two classes. For each threshold, the entropies are calculated for all intensity levels. The entropies of all the gray levels below the threshold are added and above the threshold are added separately. Final entropy expression is obtained by adding the entropy of two classes. The optimum threshold value is the maximum of all the entropy expressions. Sahoo et al. [29] used Shannon's concept of entropy, considering two probability distributions separately for foreground and background regions. The sum of the entropies of the foreground and background is then maximized.

The threshold corresponding to the maximum entropy is the optimum threshold. Sahoo et al. [29] presented a thresholding technique based on Renyi's entropy. Brink and Pendock [3], Li and Lee [17] selected a threshold by minimizing the cross entropy. Optimum threshold is then found by minimizing the cross entropy of the two regions.

Object attribute–based methods: Threshold in this method is based on property or similarity between the original image and the binarized image. These attributes may take either moment, compactness, edge matching, or stability of segmented objects. Some methods evaluate the similarity of the original image to the binary image using some fuzzy similarity measures. Tsai [33] used the moment preserving method. The objective is to choose a threshold in such a way that the mathematical moments are kept unchanged in the resulting thresholded image and the original image.

Spatial and local methods: It takes into account the correlation between pixels on a gray scale. In this class, spatial information of object and background pixels in the form of context probabilities, correlation function, co-occurrence matrices, 2-D entropy, second order statistics, etc., are considered. A criterion function is constructed using any of these parameters. Threshold is selected by minimizing/maximizing the criterion function.

Abutaleb [1] uses a two dimensional entropy, which is calculated from 2-D histogram, to get the information of the gray value of each pixel and also the average value of its immediate neighborhood. It uses spatial information to select a threshold. Thus the threshold is a vector with two entries: gray-level value of each pixel and the average value of its immediate neighborhood. Optimum threshold was obtained by maximizing the 2-D entropy. The extension of Abutaleb was also investigated by Brink [2]. He used a max–min method instead of the maximum entropic value for threshold selection, which decreases the computational complexity.

Fuzzy-based methods: It considers the uncertainty in the image due to the imprecise pixel gray levels and vagueness in image regions and boundaries. This uncertainty is incorporated in the form of membership function, which is the degree of belongingness of the pixels in an image. Then using some fuzzy measures such as entropy, fuzzy divergence, fuzzy similarity etc., an optimum threshold is selected. We now discuss some of the fuzzy-based thresholding methods in details.

5.5 Types of Fuzzy Methods

1. *Entropic based thresholding*—Shanbag [31] used an entropic based fuzzy-based thresholding method, where the membership values are computed from the cumulative probability of the gray levels. The more the gray value is away from a given threshold, the greater is the belongingness to a

specific class. Thus for any background or foreground pixels of an image, if it is i level below or above the threshold T, the membership values for foreground and background are given as:

$$\mu_f(T-i) = 0.5 + \frac{p(T) + \cdots + p(T-1-i) + p(T-i)}{2P(T)}$$

$$\mu_b(T+i) = 0.5 + \frac{p(T+1) + \cdots + p(T-1+i) + p(T+i)}{2(1-P(T))}$$

The optimum threshold is selected by minimizing the entropy

$$T_{opt} = \arg\min_{T} |H_f(T) - H_b(T)|,$$

where

$$H_f(T) = -\sum_{i=0}^{T} \frac{p(i)}{P(T)} \ln(\mu_f(i))$$

and

$$H_b(T) = -\sum_{i=T+1}^{L-1} \frac{p(i)}{1-P(T)} \ln(\mu_b(i))$$

and the probability of foreground and background area, respectively, are

$$P(T) = P_f(T) = P_f = \sum_{i=0}^{T} p(i), \quad 1 - P(T) = P_b(T) = P_b = \sum_{i=T+1}^{L-1} p(i)$$

and

$$p_i = \frac{h_i}{N}$$

where
h_i is the number of pixels of gray level, i, or the frequency of occurrence of pixel, i
N is the total number of pixels

2. *Clustering based thresholding*—Jawahar and Ray [11] thresholded the image using fuzzy clustering. Fuzzy clustering is the clustering of similar objects into a group and dissimilar objects into another group. In thresholding, the image may be clustered into two, three, or more depending on the type of objects. The problem of fuzzy clustering may be looked as partitioning a set of n sample points $X=\{x_1, x_2, x_3, \ldots, x_n\}$ into c classes such that the membership distribution has the following properties.

$\mu_j(x_i) \in [0, 1]$, $i=1, 2, 3, \ldots, n$, j=no. of classes

$$0 < \sum_{i=1}^{n} \mu_j(x_i) < n$$

and

$$\sum_{j=1}^{c} \mu_j(x_i) = 1.0$$

If h_i is the histogram and p_i is the probability of the distribution associated with the gray values of the image to be thresholded, $i \in \{0, 1, 2, 3, \ldots, L-1\}$, then the objective function

$$J = \sum_{j=1}^{2} \sum_{i=0}^{L-1} h_i \mu_j(i)^\tau d\left(i, v_j\right)^2 \tag{5.1}$$

$\tau \geq 1$ controls the fuzziness in partition and if $\tau=1$, the minima of the Equation 5.1 reduces to k-means algorithm.

The cluster means for each class

$$v_j = \frac{\sum_{i=0}^{L-1} h_i i \mu_j(i)^\tau}{\sum_{i=0}^{L-1} h_i \mu_j(i)^\tau}, \; j = 1,2 \tag{5.2}$$

characterizes the background and object region. The objective function in Equation 5.1 can be iteratively minimized by computing the means from the Equation 5.2 and updating the membership as

$$\mu_O(i) = \frac{1}{1 + \left[d(i, v_O)/d(i, v_B)^{2/(\tau-1)}\right]}$$

and

$$\mu_B(i) = 1 - \mu_O(i)$$

d represents the Euclidean distance function between gray level i and class mean.

The steps are repeated until there is no appreciable change in the μ_O and μ_B.

Formally, the algorithm to perform clustering-based image thresholding is as follows.

Algorithm 5.2: Fuzzy clustering–based image thresholding

Step 1. Initialize the membership functions of the two classes, μ_B and μ_O.
Step 2. Compute the mean gray levels of the two classes using Equation 5.2.

Step 3. Update the membership values for object and background using the formula

$$\mu_O(i) = \frac{1}{1+\left[d(i,v_O)/d(i,v_B)\right]^{2/(\tau-1)}} \quad \text{and} \quad \mu_B = 1-\mu_O.$$

Step 4. Repeat steps 2–4 until there is no appreciable change in the values μ_B and μ_O.

Example 5.1

An example in Figure 5.4 will illustrate the effectiveness of fuzzy clustering in image thresholding when partitioned in two and three classes. Figure 5.4a is a shot image (90×90) and Figure 5.4b is a thresholded image which is thresholded into three classes. Figure 5.4c is a blood image (90×90) and Figure 5.4d is a thresholded image which is thresholded into two classes.

3. *Fuzzy geometry based thresholding*—Rosenfeld [26] introduced the concept of fuzzy geometry of image subsets such as area, perimeter, and compactness. Pal and Rosenfeld [21] used the concept of fuzzy compactness to obtain an optimum threshold and standard Zadeh's *S*-function for finding the membership values of the pixels in an image. Optimum threshold is selected by minimizing the fuzzy compactness.

Consider an image A of size $M\times M$ with maximum gray level L and a_{ij} is the (i, j)th pixel of image A. For each threshold gray level, the S membership function of image A is calculated. It is defined as

$$\begin{aligned}
\mu_A(a_{ij}) &= S(a_{ij};b,c,d)\\
&= 0, \quad \text{if } a_{ij} \le b\\
&= 2\left[(a_{ij}-b/d-b)^2\right], \quad \text{if } b \le a_{ij} \le c\\
&= 1-2\left[(a_{ij}-d)/(d-b)\right]^2, \quad \text{if } c \le a_{ij} \le d\\
&= 0 \quad \text{if } a_{ij} \ge d
\end{aligned}$$

with crossover point $c=(b+d)/2$ and the bandwidth $c-b=d-c$.

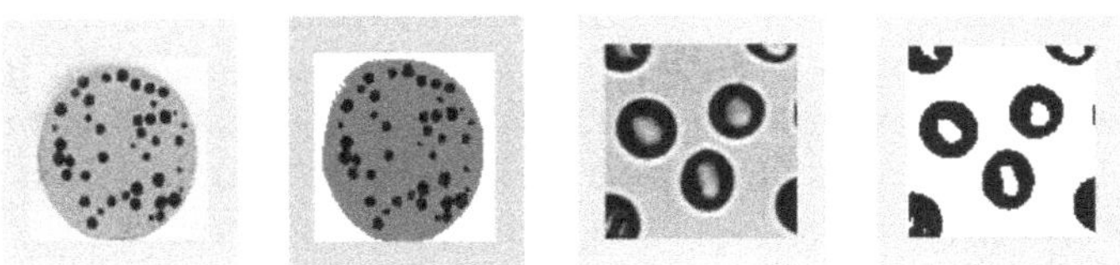

(a) (b) (c) (d)

FIGURE 5.4
(a) Shot image, (b) thresholded image (in three classes), (c) blood image, and (d) thresholded image (in two classes).

Fuzzy compactness is calculated for all threshold gray level by varying 'c' from minimum to maximum value of gray levels. Compactness is defined as follows:

Fuzzy compactness
The area and perimeter at a certain threshold 't' i.e., at 'c' = 't' is defined as

$$\text{Area: } a(\mu) = \sum_i \sum_j \mu(a_{ij}) \tag{5.3}$$

where $i, j = 1, 2, 3, \ldots, M$
Perimeter $P(\mu)$:

$$P(\mu) = \sum_{i=1}^{M} \sum_{j=1}^{M-1} \left| \mu(a_{ij}) - \mu(a_{i,j+1}) \right| + \sum_{j=1}^{M} \sum_{i=1}^{M-1} \left| \mu(a_{ij}) - \mu(a_{i+1,j}) \right| \tag{5.4}$$

The compactness of μ denoted by Comp(μ) corresponding to the threshold gray level $c = t$ is

$$\text{Comp}(\mu) = a(\mu)/p^2(\mu)$$

$$\text{Compactness}(\mu) = \text{area}(\mu)/\text{per}(\mu)^2 \tag{5.5}$$

For all threshold gray levels, fuzzy compactness is calculated and the gray level corresponding to the minimum compactness value is the threshold. When $c = t_{th}$, the image is threshold into object and background regions.

4. *Fuzzy divergence, similarity, and index of fuzziness based thresholding*—A. Chaira and Ray [6,7] proposed another fuzzy thresholding method that uses a different type of membership function called gamma membership function for finding the membership values of the pixels in an image and several fuzzy measures to select an optimum threshold. The optimum threshold is selected using the following fuzzy measures: (a) fuzzy divergence, (b) linear and quadratic indices of fuzziness, (c) fuzzy similarity, and (d) fuzzy compactness.

Consider an image A of size $M \times M$ with maximum gray level L and a_{ij} is the (i, j)th pixel of image with $i, j = 1, 2, 3, \ldots, M$. Suppose count($f$) denotes the number of occurrences of the gray level f in the image. Given a certain threshold t that separates the objects and the background regions, the average gray levels of the object region is given by the relation:

$$m_0 = \sum_{f=0}^{t} f \cdot \text{count}(f) \Big/ \sum_{f=0}^{t} \text{count}(f) \tag{5.6}$$

and the average gray level of the background region is

$$m_1 = \sum_{f=t+1}^{L-1} f \cdot \text{count}(f) \bigg/ \sum_{f=t+1}^{L-1} \text{count}(f) \tag{5.7}$$

Gamma membership function is used to find the membership function of the pixels of an image that considers the mean of the object and background regions. We now discuss briefly the gamma membership function.

5.5.1 Gamma Membership Function [7]

The general formula for the probability density function of the gamma distribution is

$$f(x) = \frac{\left(\frac{(x-b)}{\beta}\right)^{\gamma-1} \exp\left(-\frac{(x-b)}{\beta}\right)}{\Gamma(\gamma)} \quad x \geq b; \gamma, \beta > 0 \tag{5.8}$$

where
- γ is the shape parameter
- b is the location parameter
- β is the scale parameter
- Γ is the Gamma function, given by the formula

$$\Gamma(\gamma) = \int_0^{\infty} t^{\gamma-1} e^{-t} dt$$

Case 1. When $b=0$ and $\beta=1$, the Gamma distribution in Equation 5.8 becomes

$$f(x) = \frac{(x)^{\gamma-1} \exp(-x)}{\Gamma(\gamma)} \quad x \geq 0; \gamma > 0 \tag{5.9}$$

which is known as standard Gamma distribution.

Case 2. When $b \neq 0$, $\beta=1$, and $\gamma=1$, the gamma distribution in Equation 5.8 will take the form

$$f(x) = \exp(-(x-b)) \quad \text{as } \Gamma(1) = 1 \tag{5.10}$$

For the purpose of thresholding, each pixel of an image possesses a relationship to the object or background regions. If the pixel belongs to the object, it possesses a close relationship to the object region. The membership function of a pixel depends on its belongingness to a region.

For a given threshold value, the relationship between the pixel and its belonging region should depend on the absolute difference of the mean of the belonging region and the gray level. If the difference is less, the membership value of the pixel will be more and vice versa.

Replacing b in Equation 5.10 by m_0 and m_1 (Equations 5.6 through 5.7), the membership function for the object and background regions become

$$\begin{aligned}\mu_A(a_{ij}) &= \exp\left(-c\cdot|a_{ij}-m_0|\right) \quad \text{if } a_{ij}\le t, \text{ for object}\\ &= \exp\left(-c\cdot|a_{ij}-m_1|\right) \quad \text{if } a_{ij}> t, \text{ for background}\end{aligned} \tag{5.11}$$

where t is the any chosen threshold and a_{ij} is the (i, j)th pixel in the image A.

The constant c is chosen to ensure membership of the gray level representable in the range [0,1] and is calculated as $c=1/(f_{max}-f_{min})$, where f_{min} and f_{max} are the minimum and maximum gray level in the image, respectively.

For trilevel thresholding, where there are three regions in the image, two thresholds values t_1 and t_2 are selected such that $0\le t_1<t_2\le L-1$, where L is the maximum gray level of the image.

Following the same concept as in bilevel thresholding, the membership function in case of trilevel thresholding takes the form:

$$\begin{aligned}\mu_A(a_{ij}) &= \exp\left(-c_1\cdot|a_{ij}-m_0|\right) \quad \text{if } a_{ij}\le t_1\\ &= \exp\left(-c_1\cdot|a_{ij}-m_1|\right) \quad \text{if } t_1<a_{ij}\le t_2\\ &= \exp\left(-c_1\cdot|a_{ij}-m_2|\right) \quad \text{if } a_{ij}> t_2\end{aligned} \tag{5.12}$$

where m_0, m_1, and m_2 are the mean gray levels for the three regions separated by thresholds t_1 and t_2 and constant c_1 is like c to control the range of $\mu_A(a_{ij})$ in [0,1].

For selecting the optimum threshold, fuzzy measures, measures of fuzziness, and fuzzy compactness are used. We now describe the following fuzzy measures that are used to find the optimal threshold and the criterion for selecting the optimum threshold.

5.5.1.1 Fuzzy Divergence

The concept of fuzzy divergence [8] as discussed in Chapter 3 is used in image thresholding [6,7]. For two images A and B of sizes $M\times M$ with L distinct gray levels, the amount of information of $\mu_A(a_{ij})$ and $\mu_B(b_{ij})$ of images A and B at a particular point (i, j) in the image, is given by

$$e^{\mu_A(a_{ij})}/e^{\mu_B(b_{ij})} = e^{\mu_A(a_{ij})-\mu_B(b_{ij})} \tag{5.13}$$

where $\mu_A(a_{ij})$ and $\mu_B(b_{ij})$ are the membership values of the (i, j)th pixels in images A and B, respectively. $i, j = 0.1, 2, 3, \ldots, M-1$.

The discrimination of image A against image B may be given as

$$D_1(A,B) = \sum_{i=0}^{M-1}\sum_{j=0}^{M-1}\left(1-\left(1-\mu_A(a_{ij})\right)e^{\mu_A(a_{ij})-\mu_B(b_{ij})} - \mu_A(a_{ij})e^{\mu_B(b_{ij})-\mu_A(a_{ij})}\right) \quad (5.14)$$

Likewise the discrimination of B against A is

$$D_1(B,A)\sum_{i=0}^{M-1}\sum_{j=0}^{M-1}\left(1-\left(1-\mu_B(b_{ij})\right)e^{\mu_B(b_{ij})-\mu_A(a_{ij})} - \mu_B(b_{ij})e^{\mu_A(a_{ij})-\mu_B(b_{ij})}\right) \quad (5.15)$$

So, total fuzzy divergence between A and B from Equations 5.14 and 5.15 is written as

$$D(A,B) = D_1(A,B) + D_1(B,A)$$

$$\sum_{i=0}^{M-1}\sum_{j=0}^{M-1}\begin{pmatrix} 2-\left(1-\mu_A(a_{ij}) + \mu_B(b_{ij})\right)e^{\mu_A(a_{ij})-\mu_B(b_{ij})} \\ -\left(1-\mu_B(b_{ij}) + \mu_A(a_{ij})\right)e^{\mu_B(b_{ij})-\mu_A(a_{ij})} \end{pmatrix} \quad (5.16)$$

In the method, image A is an original image and image B is an ideally segmented image. An ideally segmented image is that image which is perfectly thresholded so that each pixel belongs exactly either to the object or to the background region. For an ideally thresholded image, all the object/background pixels should contribute wholly to the object/background regions, that is, the membership values of all the object pixels to the object class, that is, μ_{ij}^{obj} should be 1 and similarly the membership values of the background pixels in the background class, that is, μ_{ij}^{back} should be equal to 1.

Thus fuzzy divergence in Equation 5.16 is rewritten considering the membership values of the pixels of an ideally segmented image B, $\mu_B(b_{ij}) = 1$, as

$$D(A,B) = \sum_{i=0}^{M-1}\sum_{j=0}^{M-1}\left(2-\left(1-\mu_A(a_{ij})+1\right)\cdot e^{\mu_A(a_{ij})-1} - \left(1-1+\mu_A(b_{ij})\right)\cdot e^{1-\mu_A(a_{ij})}\right)$$

implies $$D(A,B) = \sum_{i=0}^{M-1}\sum_{j=0}^{M-1}\left(2-\left(2-\mu_A a_{ij}\right)\cdot e^{\mu_A a_{ij}-1} - \mu_A a_{ij}\cdot e^{1-\mu_A a_{ij}}\right) \quad (5.17)$$

Divergence is calculated for all threshold gray levels and the threshold corresponding to the minimum divergence value is selected as the optimum threshold.

5.5.1.2 Index of Fuzziness

The index of fuzziness by Kaufmann [13] is used in image thresholding. It denotes the degree of ambiguity present in the image by measuring the distance between membership values of the image A and its nearest ordinary set. The index of fuzziness is defined as

$$I(A) = \left(2/n^k\right) \cdot d\left(A, \tilde{A}\right) \tag{5.18}$$

where $d(A, \tilde{A})$ denotes the distance between image A and its nearest ordinary image $\tilde{A}$.

An ordinary image $\tilde{A}$ nearest to the fuzzy image A is defined as

$$\begin{aligned} \mu_{\tilde{A}}(a_{ij}) &= 0 \quad \text{if } \mu_A(a_{ij}) \le 0.5 \\ &= 1 \quad \text{if } \mu_A(a_{ij}) > 0.5 \end{aligned}$$

There are two types of indices of fuzziness: (a) linear index of fuzziness and (b) quadratic index of fuzziness. These are defined as

1. If $k=1$, d becomes the Hamming distance, then the linear index of fuzziness is

$$\begin{aligned} d\left(A, \tilde{A}\right) &= \sum_{i=0}^{M-1}\sum_{j=0}^{M-1} \left|\mu_A(a_{ij}) - \mu_{\tilde{A}}(a_{ij})\right| \\ &= \sum_{i=0}^{M-1}\sum_{j=0}^{M-1} \mu_{A \cap \bar{A}}(a_{ij}) \end{aligned} \tag{5.19}$$

$\mu_{A \cap \bar{A}}(a_{ij})$ denotes the intersection (common) of the membership value of the pixel a_{ij} at point (i, j) in the image A and its complement. So linear index of fuzziness may be rewritten as

$$\gamma_l(\mu) = \left(\frac{2}{n}\right) \sum_{i=0}^{M-1}\sum_{j=0}^{M-1} \min\left(\left(\mu_A(a_{ij})\right), \left(1 - \mu_A(a_{ij})\right)\right) \tag{5.20}$$

2. If $k=0.5$, d becomes the Euclidean distance, then the quadratic index of fuzziness is defined as

$$\gamma_q(\mu) = \left(\frac{2}{\sqrt{n}}\right) \left(\sum_{i=0}^{M-1}\sum_{j=0}^{M-1} \left(\mu_A(a_{ij}) - \mu_{\tilde{A}}(a_{ij})\right)^2\right)^{1/2}$$

$$\gamma_q(\mu) = \left(\frac{2}{\sqrt{n}}\right)\left(\sum_{i=0}^{M-1}\sum_{j=0}^{M-1}\left(\min\left(\mu_A(a_{ij}), \left(1-\mu_A(a_{ij})\right)\right)\right)^2\right)^{1/2} \quad (5.21)$$

where $n = M \times M$

In this type of fuzziness, nearest ordinary set is either 0 or 1 depending on the membership values less than or greater than 0.5. Another constant term of ln(2)=0.6931 is used to control the membership values in the range [0.5,1] so that the nearest ordinary set is set as 1. Then the membership function in Equation 5.11 is remodified as

$$\begin{aligned}\mu_A(a_{ij}) &= \exp\left(-0.6931 \cdot c \cdot |a_{ij} - m_0|\right) \quad \text{if } a_{ij} \le t, \text{ for object}\\ &= \exp\left(-0.6931 \cdot c \cdot |a_{ij} - m_1|\right) \quad \text{if } a_{ij} > t, \text{ for background}\end{aligned} \quad (5.22)$$

where t is the any chosen threshold.

Index of fuzziness is calculated for all threshold gray levels. The gray level corresponding to the minimum indices of fuzziness is selected as the threshold. The mechanism is that, for any threshold when the membership values are more toward 1, the distance from the nearest ordinary set (i.e., 1) is less, implies that the index of fuzziness is less. So when selecting the minimum value among all the values of the indices of fuzziness, implies that the separation between the calculated thresholded image and the ideally thresholded image is less.

5.5.1.3 Fuzzy Similarity Measure

This measure is proposed by Wang [34] and is used in image thresholding by Chaira and Ray [6]. The similarity between two images, A and B, may be given as

$$S_1(A, B) = \sum_{j=0}^{M-1}\sum_{i=0}^{M-1}\frac{\min\left(\mu_A(a_{ij}), \mu_B(b_{ij})\right)}{\max\left(\mu_A(a_{ij}), \mu_B(b_{ij})\right)} \quad (5.23)$$

where $\mu_A(a_{ij})$ and $\mu_B(b_{ij})$ are the membership values of the (i, j)th pixel, that is, a_{ij} and b_{ij} of the images A and B, respectively. The image B is an ideally segmented image. Similarity means that the thresholded image is similar to the ideally segmented image. So the measure is modified after substituting $\mu_B(b_{ij}) = 1$ in Equation 5.23 as

$$S_1(A, B) = \sum_{j=0}^{M-1}\sum_{i=0}^{M-1}\frac{\min\left(\mu_A(a_{ij}), 1\right)}{\max\left(\mu_A(a_{ij}), 1\right)} \quad (5.24)$$

For each threshold, the similarity values are computed. At a particular threshold value, if the similarity value between the ideally thresholded image and the calculated thresholded image is more, the segmentation is better. So the optimum threshold is the threshold value at which the fuzzy similarity measure is maximum.

Fuzzy compactness as already described in Chapter 3 is maximized to obtain an optimum threshold.

Searching strategy—For bilevel thresholding, instead of going through linear search, searching is restricted only in a zone around the valley in between the peaks as obtained from the histograms. Initially, the valleys are noted, and then the peaks just preceding and succeeding the valley are noted. The region between the two peaks is the region for searching. In this way for all the valleys, succeeding and preceding peaks are noted and the region between these two peaks is the searching region. For unimodal thresholding linear search is used from minimum to maximum value of the gray level.

B. Huang [10] minimized the fuzzy measures such as entropy (Shannon, Yager) to threshold an image. They suggested an inverse membership function $\mu_A(a_{ij})$ as

$$\begin{aligned}\mu_A(a_{ij}) &= 1/\left(1+abs(a_{ij}-m_0)^* c\right) \quad \text{if } a_{ij} \le t \\ &= 1/\left(1+abs(a_{ij}-m_1)^* c\right) \quad \text{if } a_{ij} \le t \end{aligned} \tag{5.25}$$

where m_0 and m_1 from Equations 5.6 and 5.7 are the means of foreground or background class. The constant is taken as the inverse of the difference of the maximum value and the minimum value of the image, $c=1/(f_{max}-f_{min})$.

The constant "c" is used to make the membership value of all the points lying between 0.5 and 1, that is, $0.5 \le \mu_A(a_{ij}) \le 1$. This method can be extended to multilevel thresholding.

Huang's method is also applicable to trilevel (multilevel) threshold.

Then Shannon's entropy for an image of size M X M is given as

$$E(A) = \frac{1}{M^2 \ln 2} \sum_{j=0}^{M-1} \sum_{i=0}^{M-1} -\mu_A(a_{ij}) \ln\left[\mu_A(a_{ij})\right]$$
$$-\left[1-\mu_A(a_{ij})\right] \ln\left[1-\mu_A(a_{ij})\right], \quad i,j = 0,1,2,\ldots M-1.$$

is used to obtain an optimum threshold. The entropy is calculated for all threshold gray levels. The optimum threshold corresponds to the minimum entropy value.

Another measure used is the Yager's measure. It is the measure of fuzziness between fuzzy set and its complement. The distance between a fuzzy image set and its complement is defined as

$$D(A,\bar{A}) = \left[\sum_{j=0}^{M-1} \sum_{i=0}^{M-1} |\mu_A(a_{ij}) - \mu_{\bar{A}}(a_{ij})|^p \right]^{1/p}$$

where $\mu_{\bar{A}}(a_{ij}) = 1 - \mu_A(a_{ij})$. p may be 1 or 2 for the Hamming and Euclidean distances, respectively. The measure of fuzziness is defined as:

$$\eta_p(A) = 1 - \frac{D_p(A,\bar{A})}{|A|^{\frac{1}{p}}}$$

$$\text{or } \eta_p(A) = 1 - \frac{D_p(A,\bar{A})}{|M \times M|^{\frac{1}{p}}}$$

The Yager's measure is calculated for all threshold gray levels. The optimum threshold corresponds to the minimum entropy value/measure.

The fuzzy thresholding algorithm that uses entropy or fuzzy measures is as follows.

Algorithm 5.3: Fuzzy measure/entropy-based image thresholding

Step 1. For each threshold gray level t, calculate the mean of the background and foreground regions.
Step 2. Compute the membership degree of the pixels of an image for each threshold gray level.
Step 3. Apply fuzzy measures or fuzzy entropy and calculate the fuzzy measure/entropy for each threshold gray level.
Step 4. Select the maximum/minimum value of fuzzy measure/entropy.
Step 5. Threshold the image with the threshold level corresponding to the maximum/minimum value of the fuzzy measures/entropy.

Example 5.2

Examples on fuzzy image thresholding that uses (a) fuzzy divergence, (b) measures of fuzziness, (c) fuzzy similarity, (d) compactness, and (e) fuzzy entropy for optimum threshold selection are given in Chaira and Ray [5], Huang and Wang [8]. Different types of real-time images such as blood cell, blood vessel images, shot, and other types of images will illustrate the effect of fuzzy thresholding (bilevel and multilevel) in an image. The images, namely, rice (150×150), blood cell (90×90), and shot (90×90) are in "tif" format. For bilevel thresholding, rice, blood cell images are used. Shot image is used for trilevel thresholding which contains three peaks and two valleys. Rice image, mixed with Gaussian noise, is used for unimodal thresholding. Figures 5.5 through 5.8b show the thresholded results using fuzzy divergence; Figures 5.5 through 5.7c and 5.8b show the thresholded

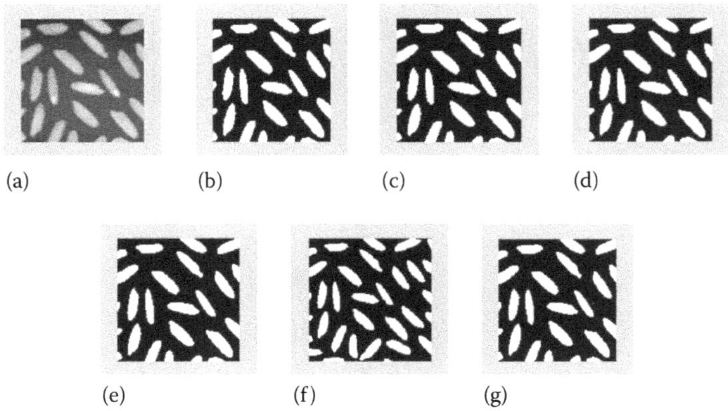

FIGURE 5.5
(a) Input image "rice," (b) thresholded image using fuzzy divergence, (c) thresholded using quadratic index, (d) thresholded using linear index, (e) thresholded using fuzzy similarity, (f) thresholded using fuzzy compactness, and (g) thresholded using Huang's method. (From Chaira, T. and Ray, A.K., *Pattern Recogn. Lett.*, 25, 865, 2004. With permission.)

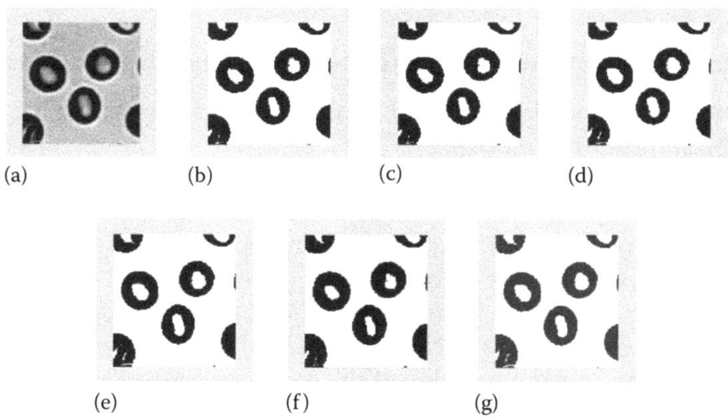

FIGURE 5.6
(a) Input image "blood," (b) thresholded image using fuzzy divergence, (c) thresholded using quadratic index, (d) thresholded using linear index, (e) thresholded using fuzzy similarity, (f) thresholded using fuzzy compactness, and (g) thresholded using Huang's method. (From Chaira, T. and Ray, A.K., *Pattern Recogn. Lett.*, 25, 865, 2004. With permission.)

images using quadratic index of fuzziness; Figures 5.5 through 5.7d and 5.8c show the thresholded images using linear index of fuzziness; Figures 5.5 through 5.7e and 5.8d show the thresholded images using fuzzy similarity; Figures 5.5 through 5.7f and 5.8e show the thresholded images using fuzzy compactness; and Figures 5.5 through 5.7g and 5.8f show the thresholded images using Huang's entropy method.

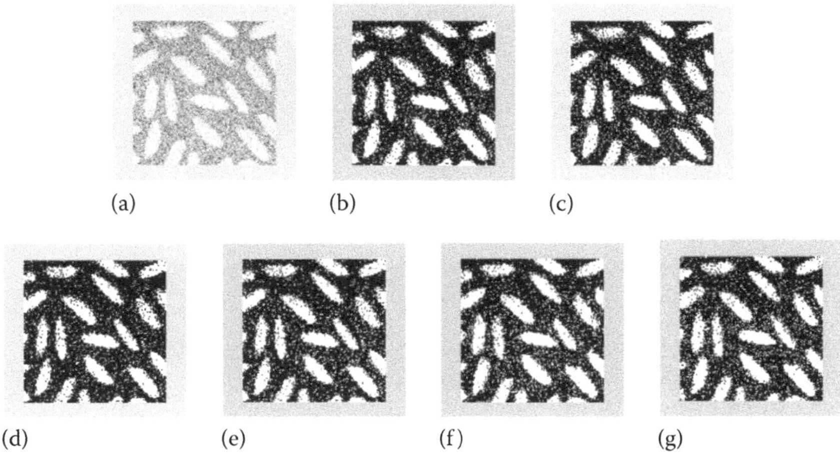

FIGURE 5.7
(a) Input image "rice" (unimodal), (b) thresholded image using fuzzy divergence, (c) thresholded using quadratic index, (d) thresholded using linear index, (e) thresholded using fuzzy similarity, (f) thresholded using fuzzy compactness, and (g) thresholded using Huang's method. (From Chaira, T. and Ray, A.K., *Pattern Recogn. Lett.*, 25, 865, 2004. With permission.)

5.6 Application of Thresholding

Image binarization is required in many industrial and medical machine vision applications. In medical imaging, there is sometimes a need to detect the changes due to the abnormal activity of the blood cells or sometimes there is a need to count the blood cells for different types of disease detection. In video surveillance, there is sometimes a need to track the activity of the object when cameras are recording around 20–30 frames per second. In these cases, thresholding is required to observe the change detection in the cell formation or the activity of the object in video. In the case of processing of document images, thresholding is suitable to separate the characters and graphics from the background.

Example 5.3: Examples on Biomedical Images

Few examples in Figures 5.9 and 5.10 on biomedical images are shown to illustrate the efficacy of fuzzy thresholding on the medical images. These are blood vessel images which are in "bmp" format. Blood vessel images are very poorly illuminated and thresholding becomes very difficult. Blood vessel thresholding is very important in pathological purpose when there is a need to count the number of blood vessels in diagnosing diseases such as angiogenesis and prostate cancer.

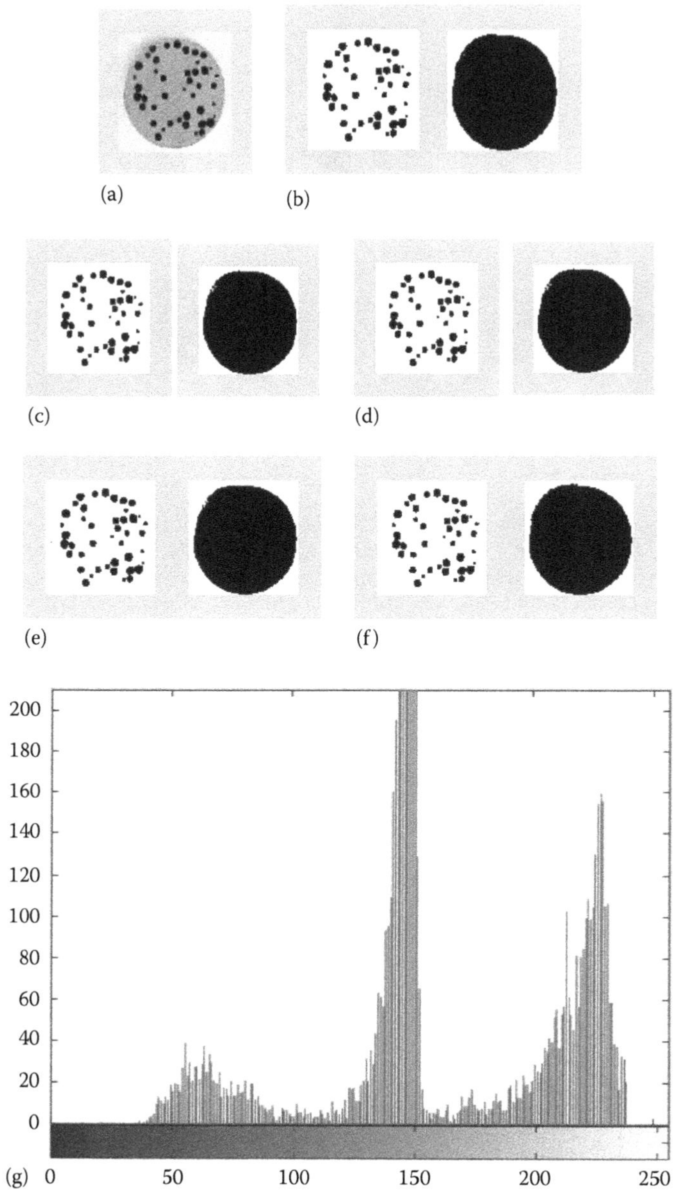

FIGURE 5.8
(a) Input image "shot," (b) thresholded image using fuzzy divergence and quadratic index, (c) thresholded using linear index, (d) thresholded using fuzzy similarity, (e) thresholded using fuzzy compactness, (f) thresholded using Huang's method, and (g) histogram of the input image. (From Chaira, T. and Ray, A.K., *Pattern Recogn. Lett.*, 25, 865, 2004. With permission.)

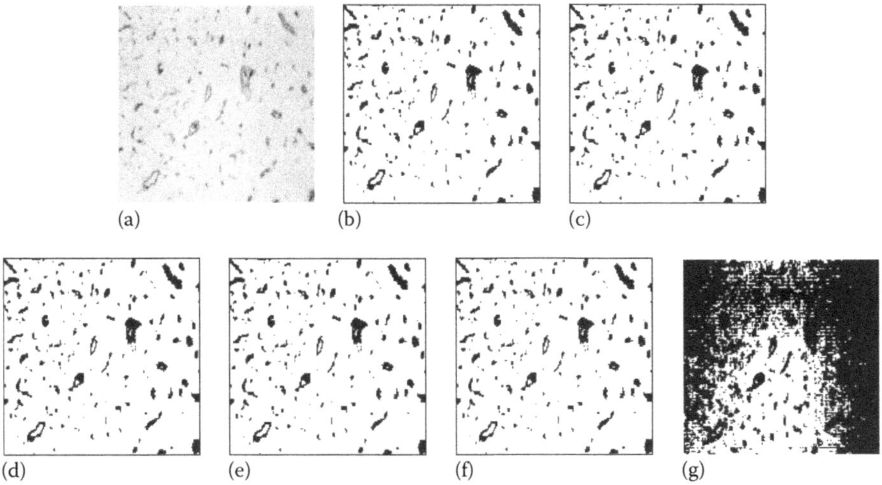

FIGURE 5.9
(a) Input image blood vessel, (b) thresholded image using fuzzy divergence, (c) thresholded using quadratic index, (d) thresholded using linear index, (e) thresholded at fuzzy similarity, (f) thresholded using fuzzy compactness, and (g) thresholded using Huang's method.

From the results in Figure 5.10, it is observed that the fuzzy algorithm that uses fuzzy divergence and quadratic index of fuzziness, fuzzy compactness give better result on poor contrasted medical images. Table 5.1 shows the thresholded values of the images for different methods.

5.7 Summary

Thresholding is a computationally simple and fast image segmentation technique where the objects are separated from the background. Thresholding may be global or local. In global thresholding, there is a single threshold value throughout the image whereas in local thresholding there are multiple threshold values in an image.

There are many threshold detection methods—P-tile threshold, mode method, histogram shape based. But, for the cases where the histogram peaks are not clear, optimal thresholding is done, where a criterion function is designed to find the separation of the regions. Thresholding of an image may be done in two ways. First, the image is considered crisp/non-fuzzy and the image pixels are considered precise and the regions of the image are also well defined. Second, the image is considered fuzzy, where the image gray levels are considered imprecise and the regions and

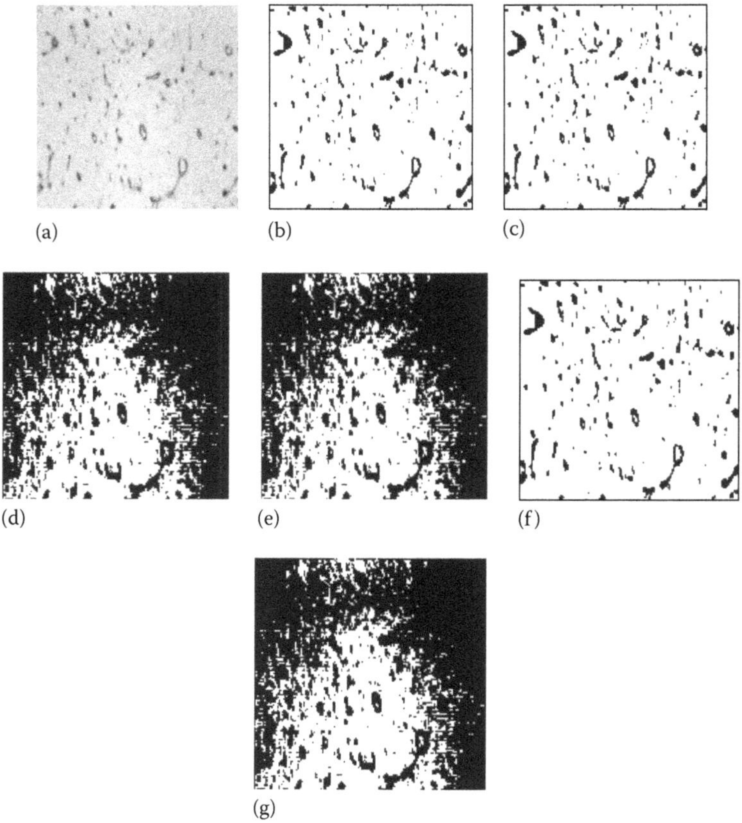

FIGURE 5.10
(a) Input image blood vessel, (b) thresholded image using fuzzy divergence, (c) thresholded using quadratic index, (d) thresholded using linear index, (e) thresholded using fuzzy similarity, (f) thresholded using fuzzy compactness, and (g) thresholded using Huang's method.

TABLE 5.1
Threshold Value for the Images Using Different Methods

Image	Divergence	Linear Index	Quadratic Index	Fuzzy Similarity	Fuzzy Compactness	Huang's Method
Rice	131, 132	131, 132	131, 132	131–132	134	131, 132
Blood	107–113	97–101	107–113	97–101	114–118	97–101
Cameraman	88–89	75	88–89	75	90	64
Shot	106, 180	108, 184	106, 180	106, 184	110, 180	104, 186
Rice (unimodal)	179	179	177–178	179	174	175
Blood vessel	172	173	172	173	173	201
Blood vessel 1	163	192	163	192	167	192

Source: Chaira, T. and Ray, A.K., *Pattern Recogn. Lett.*, 25, 865, 2004. With permission.

boundaries are considered vague. The fuzziness is incorporated by introducing membership degree to each pixel of the image. It utilizes the fuzzy methods of an input image to identify an appropriate threshold value of the image. Fuzzy thresholding is performed using different types of fuzzy measures. Various fuzzy thresholding methods are discussed and few examples are shown on real-time images as well as on medical images to illustrate the efficacy of fuzzy thresholding methods. The use of fuzzy range can help to locate effectively the deep valley in the histogram.

References

1. Abutaleb, A.S., Automatic thresholding of gray-level pictures using two dimensional entropy, *Computer Vision Graphics and Image Processing*, 47, 22–32, 1989.
2. Brink, A.D., Thresholding of digital images using two dimensional histogram and fuzzy entropy principle, *IEEE Transactions in Image Processing*, 9, 4, 2000.
3. Brink A.D. and Pendock, N.E., Minimum cross-entrophy threshold selection. *Pattern Recognition*, 29, 179–188, 1996.
4. Chang, C., Chen, K., Wang, J., and Althouse, M.L.G., A relative entropy based approach in image thresholding, *Pattern Recognition*, 27, 1275–1289, 1994.
5. Cheng, H.D., Chen, Y.H., and Sun, Y., A novel fuzzy entropy approach to image enhancement and thresholding, *Signal Processing*, 75, 277–310, 1999.
6. Chaira, T. and Ray, A.K., Threshold selection using fuzzy set theory, *Pattern Recognition Letters*, 25, 865–874, 2004.
7. Chaira, T. and Ray, A.K., Segmentation using fuzzy divergence, *Pattern Recognition Letters*, 12, 24, 1837–1844, 2003.
8. Fan, J. and Xie, W., Distance measure and induced fuzzy entropy, *Fuzzy Sets and Systems* 104, 305–314, 1999.
9. Gonzalez, R.C. and Woods, R.E., *Digital Image Processing 2nd Ed.*, Prentice Hall, Upper Saddle River, NJ, 2002.
10. Huang, L.K. and Wang, M.J.J., Image thresholding by minimizing the measure of fuzziness, *Pattern Recognition*, 28, 41–51, 1995.
11. Jawahar, C.V. and Ray, A.K., Investigations on fuzzy thresholding based on fuzzy clustering, *Pattern Recognition*, 30(10), 1605–1613, 1997.
12. Kapur, J.N., Sahoo, P.K., and Wong, A.K.C., A new method for gray level picture threshold using the entropy of the histogram, *Graphical Models and Image Processing*, 29, 273–285, 1985.
13. Kaufmann, A., *Introduction to the Theory of Fuzzy Subsets: Fundamental Theoretical Elements*, Vol. 1, Academic Press, New York, 1980.
14. Kittler, J. and Illingworth, J., On threshold selection using clustering criteria, *IEEE Transactions on System, Man and Cyber*netics, SMC-15, 652–655, 1985.
15. Kittler, J. and Illingworth, J., Minimum error thresholding, *Pattern Recognition*, 19, 41–47, 1986.
16. Leung, C.K., Performance analysis for a class of iterative image thresholding algorithms, *Pattern Recognition*, 29(9), 1523–1630, 1996.

17. Li, E.H. and Lee, C.K., Minimum cross-entropy thresholding, *Pattern Recognition*, 26(4), 617–625, 1992.
18. Niblack, W., *An Introduction to Image Processing*, Prentice-Hall, Englewood Cliffs, NJ, 1986.
19. Otsu, N., A threshold selection method from gray level histograms, *IEEE Transactions on Systems, Man, and Cybernetics*, SMC-9, 62–66, 1979.
20. Pal, N.R., On minimum cross-entropy thresholding, *Pattern Recognition*, 29(4), 575–580, 1996.
21. Pal, S.K. and Dasgupta, A., Spectral fuzzy sets and soft thresholding, *Information Sciences*, 654, 65–97, 1992.
22. Pal, N.R. and Pal, S.K., Entropy, a new definition and its application, *IEEE Proceedings*, 36, 284–295, 1989.
23. Pal, S.K. and Rosenfeld, A., Image enhancement and thresholding by optimization of fuzzy compactness, *Pattern Recognition Letters*, 7, 77–86, 1988.
24. Ramesh, N., Yoo, J.H., and Sethi, I.K., Thresholding based on histogram approximation, *IEE Proceedings on Vision Image and Signal Processing*, 142(5), 271–279, 1995.
25. Ridler, T., and Calvard, S., Picture thresholding using an iterative selection method (1978), *IEEE Transactions on Systems, Man, and Cybernetics*, 20(5), 630-632, 1978.
26. Rosenfeld, A., The fuzzy geometry of image subsets, *Pattern Recognition Letters*, 2, 311–317, 1984.
27. Rosenfeld, A. and Torre, P., Histogram concavity analysis as an aid in threshold selection, *IEEE Transactions on System, Man and Cybernetics*, SMC-13, 231–235, 1983.
28. Sahoo, P.K., Soltani, S., and Wong, A.K.C., A survey of thresholding techniques, *Computer Vision, Graphics and Image Processing*, 15, 319–338, 1981.
29. Sahoo, P.K., Wilkins, C., Yeager, J., Threshold selection using Renyi's entropy, *Pattern Recognition*, 30, 71–84, 1997.
30. Sankur, B. and Sezgin, M., Survey over image thresholding techniques and quantitative performance evaluation, *Journal of Electronics Imaging*, 13(1), 146–165, 2004.
31. Shanbag, A.G., Utilization of information measure as a means of image thresholding, *Computer Vision Graphics and Image Processing*, 56, 414–419, 1994.
32. Sonka, M. et al., *Image Processing, Analysis, and Machine Vision*, Brooks/Cole, Pacific Grove, CA.
33. Tsai, W.H., Moment preserving thresholding: A new approach, *Graphical Models and Image Processing*, 19, 377–393, 1985.
34. Wang, W.J., New similarity measures on fuzzy sets and fuzzy elements, *Fuzzy Sets and Systems*, 85, 305–309, 1997.
35. Whatmough, R.J., Automatic threshold selection from a histogram using the exponential hull, *Graphical Models and Image Processing*, 53, 592–600, 1991.
36. Wong, A.K.C. and Sahoo, P.K., A gray level threshold selection method on maximum entropy principle, *IEEE Transactions on System, Man and Cybernetics*, SMC-19, 866–871, 1989.

6

Fuzzy Match–Based Region Extraction

6.1 Match-Based Region Extraction

In many image-processing applications, an important task is to locate a set of known objects, patterns, or regions in an image. Thus given an image and a set of objects of interest, it is important to identify and locate a given set of objects or regions in the image. The problem is one of segmenting an image and extracting the regions of interest using matching techniques.

Match-based object/region extraction thus essentially finds the correspondence or the correlation between each pair of homologous image points on a given pair of images. Matching is performed by comparing a query object or sub-image region with the test image to find the location of the matched pattern in the scene image. The procedure thus involves comparing the gray-level properties or other properties, such as texture, of the query object region with the same properties of the image. The query object or sub-image pattern may or may not be a part of the scene image.

The technique may be employed on either stationary or dynamic images. Region extraction using match-based segmentation yields excellent results if an exact pattern or region of interest is present in the scene image. The aim is to obtain the spatial boundaries of the region that is of interest to the user. Matching criteria requires the following:

1. Evaluating the match criteria for each location in the image
2. Calculating the local maxima of this criterion
3. Searching for a pattern location, if this maximum exceeds a predefined threshold

Match-based segmentation localizes all image positions at which close copies of search pattern are located. The object pattern may present anywhere in the image and all the regions corresponding to the regions are extracted, as shown in Figure 6.1.

The query object/regions may present at multiple locations in the scene image. Thus given an image I and a model image $\Im$, we are to extract the model image $\Im$ in the scene image, *I*. The model patterns may use any combination of features like gray-level intensity, texture, and color. When a set of model images is used to extract the regions of different patterns in the scene image, *I*, all the regions of similar patterns or models, present in

(a) (b) (c)

FIGURE 6.1
(a) Original image, (b) query image, and (c) extracted region as that of the query image in (b).

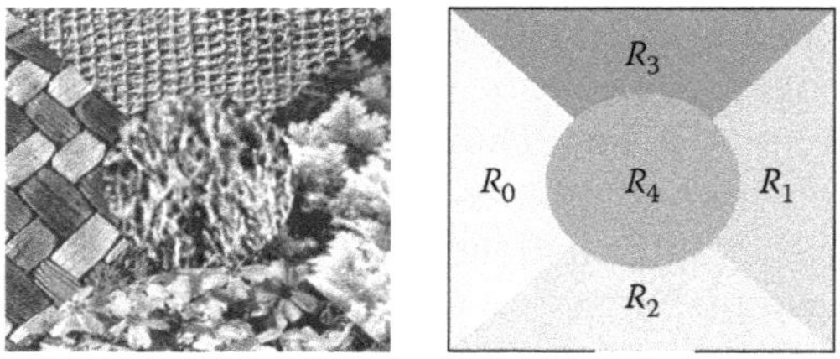

(a) (b)

FIGURE 6.2
(a) Image having five types of regions. (b) Image segmentation.

the scene image, are extracted and in this case, the problem resembles a region extraction: A set of regions, say R_i, is extracted from the image. In this case, union of all the extracted regions is not equal to 1, as all the models/patterns may not present in the scene image.

However, the difference between the match-based region extraction to that of region segmentation is that in region segmentation the criteria, that is, $\cup R_i = 1$ and $\cap R_i = 0$ follows while in region extraction, the criteria, that is, $\cup R_i \leq 1$ and $\cap R_i \geq 0$ follows. In the case of region segmentation, no such patterns are required to segment an image. The objective of segmentation is to segment all the homogeneous regions in an image to satisfy the criteria as shown in Figure 6.2. The figure shows a segmentation procedure that segments the image into five non-overlapping regions. However, the general problem of image segmentation of unconstrained images is poorly defined. When an image is not well defined, that is, when the number of regions or classes is not properly defined, then the segmentation problem may become critical. Figure 6.3 shows the two ways one can segment an image (a "Lena" image). The image does not contain clear regions and so one may use two or three ways to partition the image. There are multiple ways to choose the partitions R_i in the image. Thus, if the number of classes is not properly defined then it is not possible to establish a ground-truth segmentation in order to evaluate the correctness of any segmentation. Thus, in case of

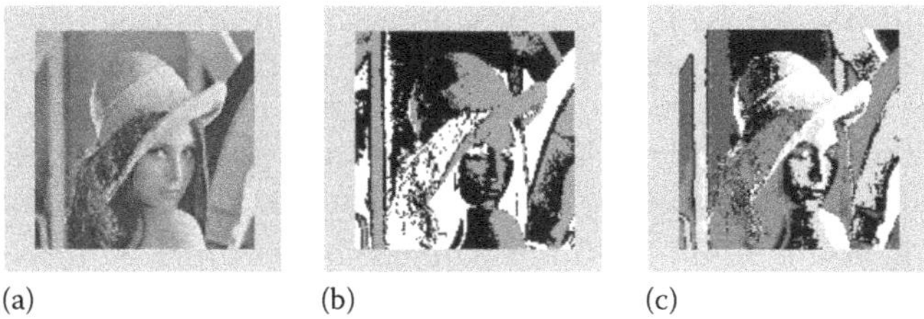

FIGURE 6.3
(a) Original "Lena" image, (b) segmented into three regions, and (c) segmented into five regions.

unconstrained images, region extraction or match-based segmentation may be useful where the criterion function $\cup R_i \leq 1$ and $\cap R_i \geq 0$ holds, that is, the extracted regions may overlap or may not completely partition the image.

There are many possible ways for region extraction. People evaluate the images, and the information is identified visually. However, manual extraction of regions and objects is very tedious and time consuming when there is large image collection. In match-based segmentation or region extraction, segmentation is done using matching criteria. So region extraction is nothing but extracting the regions of an image similar to the query or model image provided using matching criteria with image-derived features or gray-level properties.

In this chapter, fuzzy methods for match-based region extraction have been presented. The procedure for region extraction has been implemented on both monochrome and color images. It uses histogram back projection algorithm and fuzzy measures for extracting the regions in the processed image. Back projection algorithm serves as a powerful technique for extracting regions in an unconstrained image similar to the model image provided and the model image, which may appear in multiple locations in image.

6.2 Back Projection Algorithm

In this section, we present several back projection algorithms that have been widely used. These are the set of algorithms, which extract and locate within images a set of regions with feature histograms that are similar to the given model histogram [10].

6.2.1 Swain and Ballard's Back Projection Algorithm

The novelty of Swain and Ballard's back projection algorithm resides in the use of color information instead of spatial information. In Swain and

Ballard's formulation, histogram back projection determines the most likely location of the query or model histogram h_q within the scene image $I(x, y)$. Each model is slided over the scene image and the likelihood ratio of the window of the scene image where the model was placed to the model image, is computed. For a given query or model histogram h_q and the image or the target histogram h_i, if $h_r[m] = \min(h_q[m]/h_i[m], 1)$, $m \in 0, 1, 2, \ldots, M-1$, m is the gray level of an image then, each point in the image is replaced by the corresponding confidence score, $I'[x, y] = h_r[k]$, where $k = I[x, y]$. $I'(x, y)$ is a new matrix of same size as that of image. The new matrix thus represents how much it characterizes the searched objects. The image I' is convolved with a blurring mask to reduce the noise and the peak value is the expected location of the model in the image.

The back projection algorithm requires

1. An object or model histogram
2. Image or scene histogram

We are now going to cover a few non-fuzzy/crisp methods based on back projection algorithm for region extraction to have a clear idea on the use of back projection.

Enneser and Medioni [6] used the weighted histogram intersection method in much the same manner as Swain and Ballard for finding the "Waldo" in the image. If L_i and M_i be the counts in a histogram bin "i" for local histogram and the model histogram, respectively, then the formula $\sum_i W_i \min(L_i, M_i)$, where W is the weight chosen as M/I, I is the histogram of the entire image. The following Sections 6.2.2–6.2.5 non-fuzzy back projection methods are suggested by Smith [7–9].

6.2.2 Quadratic Confidence Back Projection

In quadratic confidence back projection, the ratio h_r is computed by taking the ratio of the model histogram h_q to the target histogram h_i which is given as follows:

$$h_r[m] = \frac{\sum_n h_q[m] A_{m,n}}{h_i[m]}$$

where $A_{m,n}$ is the similarity matrix that defines the similarity of the histogram elements indexed by m and n.

Each point in an image is then assigned as follows:

$$\text{If } k = I(x,y), \quad k \text{ is an index of the feature at image point } I(x,y), \text{ then } I'(x,y) = h_r[k]$$

The image $I'(x, y)$ is then smoothed to reduce noise and then thresholded for locating the best location of the model image.

6.2.3 Local Histogramming

In local histogramming, both the smoothing operation and the back projection steps are combined.

Here at each location in the target image, a local neighborhood histogram h_l is calculated and its distance (any distance metric) from the model histogram, h_q, is computed to see whether each point belongs to the model. If the dimension of the small local region R_{xy}^{wh} is $w \times h$ and its local histogram is h_l^{xy} in image $I(x, y)$, then at point (x, y)

$$I'[x,y] = d_{q,l}\left(h_q, h_l^{xy}\right), \quad \text{where } d_{q,l} \text{ is a distance metric.}$$

6.2.4 Binary Set Back Projection

In binary set back projection, the back projection of binary feature set onto the image produces a bilevel image. The level corresponds to the selected or unselected pixels, i.e., 0 or 1, of the specified model binary set s_q [7]. The binary set is equivalent to a thresholded histogram, where each bin is specified in two levels.

Let $k = I(x, y)$, then

$$I'(x,y) = s_q(k)$$

$I'(x, y)$ is the back projection of the model binary set s_q. The image $I'(x, y)$ is then filtered to remove noise and thresholded to obtain the likely locations of the model image.

6.2.5 Single Element Quadratic Back Projection

It automatically extracts the regions of an image. It is a combination of quadratic and binary set back projection methods. The method first generates color histogram $h[m]$ for each image. Then for each m, such that $h(m) > \tau$, a binary set s_q is generated such that $s_q[m] = 1$ and $\forall_{k \neq m} s_q[k] = 0$. Each binary set s_q, weighted by a similarity, is back projected onto the image such that

$$\text{if } k = I(x,y) \quad \text{then } I'(x,y) = A_{m,k}$$

where $A_{m,k}$ is the similarity of histogram elements indexed by m and k.

The image $I'(x, y)$ is then smoothed to reduce noise and thresholded to reveal the likely locations of the model image.

6.3 Fuzzy Region Extraction Methods

This section provides a method for fuzzy region extraction [2–5] on both monochrome and color images. The image is considered to be fuzzy as most of the image regions are not clear. Thus, fuzzy set theory may help a lot by incorporating the vagueness in the form of a membership degree. The region extraction method that we are going to discuss uses back projection algorithm, which is based on fuzzy similarity measures. These fuzzy measures are very useful in finding an exact match of the model image in the scene image. Before describing the fuzzy region extraction algorithm, we will first discuss about the fuzzy measures such as (a) min–max ratio, (b) normalized absolute difference, and (c) Generalized Tversky's index (GTI) model in Section 6.3.1.

6.3.1 Fuzzy Similarity Measures

For image retrieval, the similarity between two sets of features extracted from the scene image and the query image is used as a match measure. The match measure is used to retrieve those regions present in the scene image, which are similar to the query image. A particularly intuitive similarity function S, which reflects the degree of resemblance between the query image and the scene image is presented. Considering images as fuzzy sets, the similarity between two images A and B computed from the histograms A_n and B_n, $n=0, 1, 2, \ldots, L-1$, L being the maximum gray level of the image, may be given as

1. *Similarity measure based on min–max ratio*

   This similarity measure is described in Chapter 3. Chaira and Ray [2,3] used this similarity measure in region extraction.

$$S_1(A, B)=(1/L)^* \sum_{n=1}^{L} \frac{\min\left(\mu_A(n), \mu_B(n)\right)}{\max\left(\mu_A(n), \mu_B(n)\right)} \tag{6.1}$$

   where $\mu_A(n)$ and $\mu_B(n)$ are the membership values of the nth gray levels of the histograms A_n and B_n of the two images A and B, respectively. For an exactly identical pair of images, the similarity value is equal to 1.

2. *Similarity measure based on the normalized absolute difference*

   This measure, described in Chapter 3, is used by Chaira and Ray [2,3] in image region extraction. It associates with the membership values of the gray levels computed from the histograms of the two images. This measure is also observed to yield appropriately extracted regions from the scene image and is given below:

$$S_2(A,B) = 1 - \frac{\sum_{n=1}^{L} \text{abs}\left(\mu_A(n) - \mu_B(n)\right)}{\sum_{n=1}^{L} \left(\mu_A(n) + \mu_B(n)\right)} \tag{6.2}$$

For both the similarity measures, the following properties hold.

Property 1: The similarity measure $S(A,B)$ between two images A and B is symmetric, that is,

$S(A,B) = S(B,A)$ for all A_n, $B_n \in [0,1]$.

Property 2: $S(A,B) = 1$, if $A_n = B_n$.

Property 3: If $A_n \subseteq B_n \subseteq C_n$, for the histograms A_n, B_n, and C_n, $n = \{0, 1, 2, \ldots, L-1\}$ of three images A, B, and C, respectively, then

$S(A,B) \geq S(A,C)$ and $S(B,C) \geq S(A,C)$.

To prove property 3, let us consider three histograms A_n, B_n, and C_n of the images A, B, and C. $\mu_A(n)$, $\mu_B(n)$, $\mu_C(n)$ are the membership values of the nth gray levels of the histograms of the images, respectively. Also consider that the three histograms are such that $A_n \subseteq B_n \subseteq C_n$. Then we have $\mu_A(n) \leq \mu_B(n) \leq \mu_C(n)$. For Equation 6.1

$$S(A,B) = \frac{\min\left(\mu_A(n), \mu_B(n)\right)}{\max\left(\mu_A(n), \mu_B(n)\right)} = \frac{\mu_A(n)}{\mu_B(n)} \quad \text{as } \mu_A(n) \leq \mu_B(n)$$

$$\text{Similarly } S(A,C) = \frac{\min\left(\mu_A(n), \mu_C(n)\right)}{\max\left(\mu_A(n), \mu_C(n)\right)} = \frac{\mu_A(n)}{\mu_C(n)} \quad \text{as } \mu_A(n) \leq \mu_C(n)$$

and

$$S(B,C) = \frac{\min\left(\mu_B(n), \mu_C(n)\right)}{\max\left(\mu_B(n), \mu_C(n)\right)} = \frac{\mu_B(n)}{\mu_C(n)} \quad \text{as } \mu_B(n) \leq \mu_C(n)$$

Therefore, from the above three notations we have the following:

$\frac{\mu_B(n)}{\mu_C(n)}$ may be written as $\frac{\mu_A(n)}{\mu_C(n)} + \frac{\mu_B(n) - \mu_A(n)}{\mu_C(n)}$ which is greater than $\frac{\mu_A(n)}{\mu_C(n)}$.

$$\text{i.e., as } \frac{\mu_A(n)}{\mu_C(n)} \leq \frac{\mu_B(n)}{\mu_C(n)}.$$

$$\text{Likewise } \frac{\mu_A(n)}{\mu_C(n)} \leq \frac{\mu_A(n)}{\mu_B(n)}.$$

This leads to the relation $S(B,C) \geq S(A,C)$ and $S(A,B) \geq S(A,C)$.

3. *Measure based on Tversky's model*

   This measure is described in Chapter 3. Generalized Tversky's model (GTI) [11,12], which is an extension of Tversky's model that uses fuzzy features, is defined as

$$\text{GTI}\,(A, B; \alpha, \beta) = \frac{f\,(A_n \cap B_n)}{f\,(A_n \cap B_n) + \alpha \cdot f\,(A_n - B_n) + \beta \cdot f\,(B_n - A_n)} \tag{6.3}$$

   where

   A_n, B_n are the histograms of two images A and B, on the measurements and α, $\beta \geq 0$

   $A_n \cap B_n$ denotes the common features, that is, features that are common to both images A and B

   $A_n - B_n$ denotes the distinctive features that are present in image A and not in image B

Using min t-norm, Tversky Index-Min from Equation 6.3 is defined as

$$\text{TIM}(A,B;\alpha,\beta) = \frac{\sum_{n=1}^{L} \min(\mu_A(n), \mu_B(n))}{\sum_{n=1}^{L} \left(\min(\mu_A(n), \mu_B(n)) + \alpha \cdot \min(\mu_A(n), 1 - \mu_B(n)) + \beta \cdot \min(1 - \mu_A(n), \mu_B(n))\right)} \tag{6.4}$$

where $\mu_A(n)$ and $\mu_B(n)$ are the membership functions of the nth gray level of the histogram of the images A and B. When the membership values of the gray levels of the histograms are similar, the common features of the two images are more and the distinctive feature is less. This in turn implies that the two images are similar. Chaira and Ray [2,3] extended the Tversky Index-Min to find the similarity among the gray/color images. In this fuzzy method, the values of α and β are chosen as 0.5.

6.3.2 Fuzzy Measures in Region Extraction

We will discuss the algorithm for fuzzy region extraction. Given a scene image and a query image, the query image is slided over the scene image to find out the regions in the scene image where the query image is matched. Window of the scene image is the region in the scene image where the query image is placed and thus the size of the window image is same as that of the query image. The membership function is computed by normalizing the histogram of the query image and the histograms of the nonoverlapping windows in the scene image by a normalizing factor (norm). Query image is a part of the scene image and the windows are the images formed when the query image is made to slide over a scene image. The normalizing factor is calculated by taking the maximum count of the histograms of the query image and the windows of the scene image. Mathematically, the normalizing factor (norm) may be represented as

$$\text{norm} = \max(h_{\text{q}}, z) \tag{6.5}$$

where

$z = \max_r\left[\max(h_{tr})\right]$, $r=1, 2, 3, \ldots, r, \ldots, R$, is the total number of windows in the image

h_{q} and h_{tr} are the histograms of the query image and the rth window in the scene image

Fuzzy ratio is then computed using fuzzy similarity measures from Equations 6.1, 6.2, and 6.4. This fuzzy ratio is the likelihood ratio of the window of the scene image to the model image.

For each gray level, the fuzzy ratio $f(n)$ is calculated from the following similarity measures:

$$f(n) = \frac{\min\left(\mu_{\text{q}}(n), \mu_{\text{t}}(n)\right)}{\max\left(\mu_{\text{q}}(n), \mu_{\text{t}}(n)\right)}. \tag{6.6}$$

$$f(n) = 1 - \frac{abs\left(\mu_{\text{q}}(n) - \mu_{\text{t}}(n)\right)}{\left(\mu_{\text{q}}(n) + \mu_{\text{t}}(n)\right)} \tag{6.7}$$

$$f(n) = \frac{\min\left(\mu_{\text{q}}(n), \mu_{\text{t}}(n)\right)}{\left(\min\left(\mu_{\text{q}}(n), \mu_{\text{t}}(n)\right) + \alpha \cdot \min\left(\mu_{\text{q}}(n), 1 - \mu_{\text{t}}(n)\right) + \beta \cdot \min\left(1 - \mu_{\text{q}}(n), \mu_{\text{t}}(n)\right)\right)} \tag{6.8}$$

where $\mu_{\text{q}}(n)$ and $\mu_{\text{t}}(n)$ are membership values of the nth gray level of the histograms of the query and the window in the scene image as the query image is made to slide over the scene image, respectively, with "n," $n=\{0, 1, 2, \ldots, L-1\}$, representing the gray values of the image. Then each pixel gray level "n" in the image window of the scene image is replaced by the fuzzy ratio $f(n)$ of that pixel gray level.

If each point in the image is assigned as

$$I(i, j) = n, \quad \text{then } I'(i, j) = f(n)$$

After replacing the image with the fuzzy ratio value, a new image I' is formed. It is thresholded at the largest peak and filtered with Gaussian filter to reduce noise. The continuous white portion of the image is the region similar to the model image. The algorithm that summarizes the fuzzy region extraction method is given in Algorithm 6.1.

Algorithm 6.1: Fuzzy back projection algorithm

Step 1. Slide the model or query image over the scene image.
Step 2. Compute the histograms of the model image and the windows of the scene image.
Step 3. Form a likelihood ratio that the window of the scene image belongs to the model using the fuzzy ratio in Equations 6.6 through 6.8.
Step 4. Back project the fuzzy ratio on to the image so that the image values are replaced by the values of $f(n)$, which are the indices.
Step 5. Threshold the back projected image at its highest peak.
Step 6. Filter the thresholded image with Gaussian filter to obtain a region-extracted image.

Example 6.1

An example is shown [2–5] that illustrates the effect of fuzzy method for region extraction in monochrome images. Different types of Brodatz images [1] are combined to form a multi-textured database image. Each image in the scene image contains several regions and the query image or the pattern is selected from any part of the database image. In each of the resultant image, the scene image, model or query image, and the region-extracted images are shown. Results using the three similarity measures are almost similar and so the results using the similarity measure in Equation 6.1 are shown.

Figure 6.4a is a multi-textured image of size 200×200 containing four types of textured regions. Two types of query images are selected from the image. Figure 6.4b and e are the query images, which are a part of the scene image of sizes 40×40 and 20×20, respectively. Figure 6.4c and f are the region-extracted images (continuous white portion) similar to the query image. Figure 6.4d and g are the region-extracted images filtered with Gaussian filter to remove noise.

Example 6.2

Figure 6.5a is a complex multi-textured image of size 200×200 containing different types of textured regions. Figure 6.5b is the query image, which are a part of the scene image of size 40×40 which is present at many places in the input image. Figure 6.5c is the region-extracted image (continuous white portion) similar to the query image. Figure 6.5d is the region-extracted images filtered with Gaussian filter to remove noise.

Example 6.3

As it is histogram based, this process is rotation invariant up to 50°. Figure 6.6 shows the results on an image with three types of regions and the query image

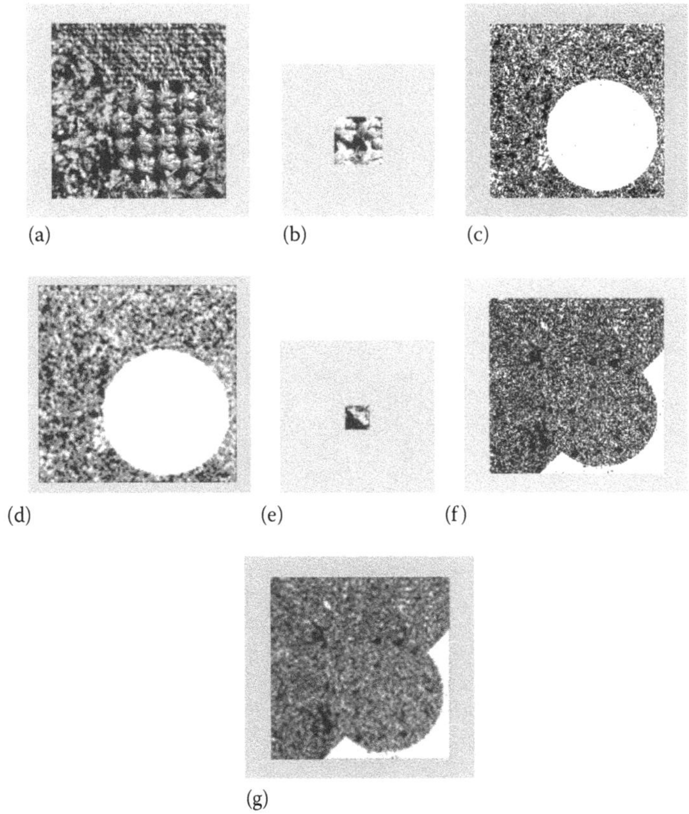

FIGURE 6.4
(a) Input image with four textured regions, (b) query image, (c) extracted region (white) as equivalent to the query image shown by arrow, (d) filtered extracted region, (e) query image, (f) extracted region as that of query image (e), and (g) filtered image of (f). (From Chaira, T. and Ray, A.K., *Int. J. Fuzzy Math.*, 11, 601, 2003. With permission.)

is rotated up to 50° as shown in Figure 6.6b. The continuous white portion after applying Gaussian filter in Figure 6.6d is the extracted region.

Similar to the monochrome images, region-extraction procedure applied on multicolored images having different color regions is provided by Chaira and Ray [2]. The membership function for each color channel is computed by normalizing the histogram of the query image, which is a part of scene image, and the histograms of the nonoverlapping window in the scene image by a normalizing factor in Equation 6.5.

Each pixel gray level "n" in the image window of the scene image is replaced by the fuzzy value $f(n)$ of the pixel gray level of the color channel. If each point in the image is assigned as

$$I(i,j) = n, \quad \text{then } I'(i,j) = f(n)$$

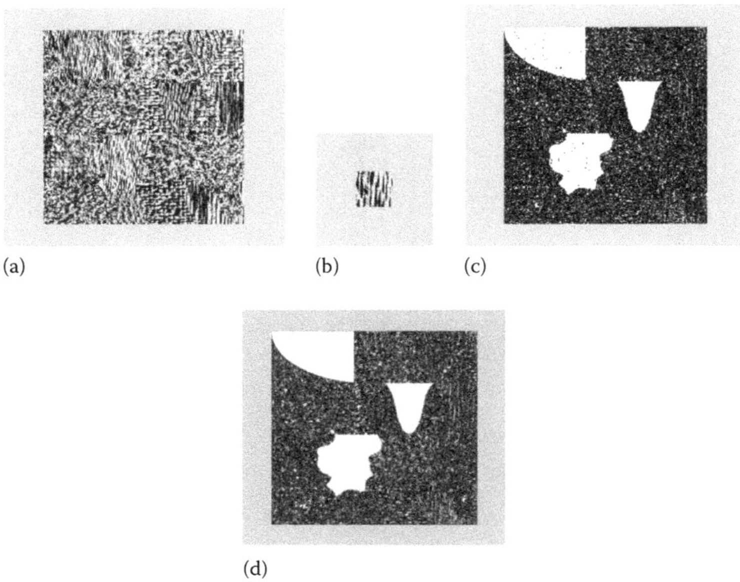

FIGURE 6.5
(a) Input image with multi-textured regions, (b) query image, (c) extracted region (white) as equivalent to the query image, and (d) filtered extracted region. (From Chaira, T. and Ray, A.K., *Int. J. Fuzzy Math.*, 11, 601, 2003. With permission.)

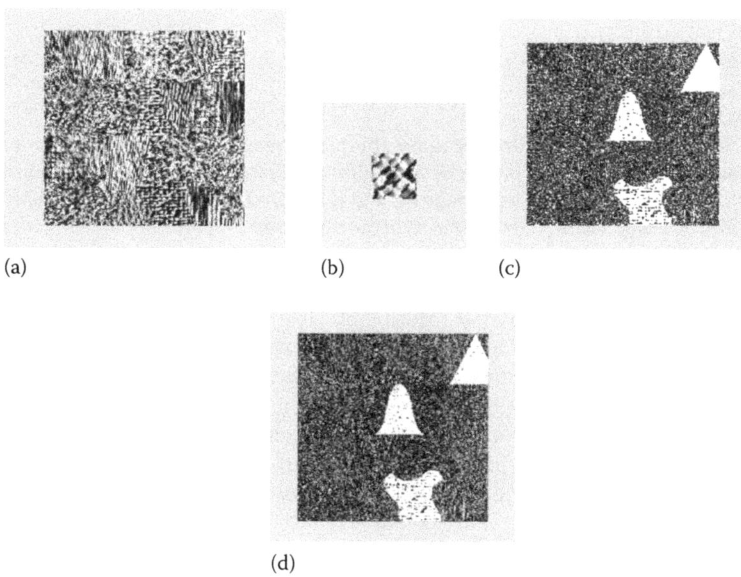

FIGURE 6.6
(a) Input image four textured regions, (b) query image (rotated), (c) extracted region (white) as equivalent to the query image, and (d) filtered extracted region (continuous white region). (From Chaira, T. and Ray, A.K., *Int. J. Fuzzy Math.*, 11, 601, 2003. With permission.)

For all the three color channels, *f*(*n*) is computed using Equations 6.6 through 6.8. After replacing the image with the fuzzy ratio value, three new images I'_R, I'_G, and I'_B are formed for three color channels, *R*, *G*, and *B*, respectively. The three new images I'_R, I'_G, and I'_B, are thresholded at the largest peak. The resulting three thresholded images are added, normalized, and again thresholded at the largest peak. The final image is filtered with Gaussian filter and displayed. The continuous white portion of the image is the region similar to the model image. In the final image, the continuous white portion is the exact extracted region.

Region extraction is very common and useful in remote sensing application. A query image may be a piece of agricultural or volcanic or watery or any other type of region and this query image is used to extract the regions that correspond to the region in the large scene image. Few results are shown on several colored scene images that uses fuzzy method by Chaira and Ray [2] and the nonfuzzy method [7]. The region-extraction results are almost similar for all the fuzzy similarity measures.

Example 6.4

Figure 6.7a shows a multicolored "zebra" image of size 340×340. This image is adapted from Smith's thesis [7]. Figure 6.7b is the query image of size 20×20, which is a part of the "zebra" image. Figure 6.7c shows the extracted region similar to the query or model image and Figure 6.7d is the filtered image with Gaussian filter of the Figure 6.7c. Figure 6.7e shows the region-extracted image using Smith's method.

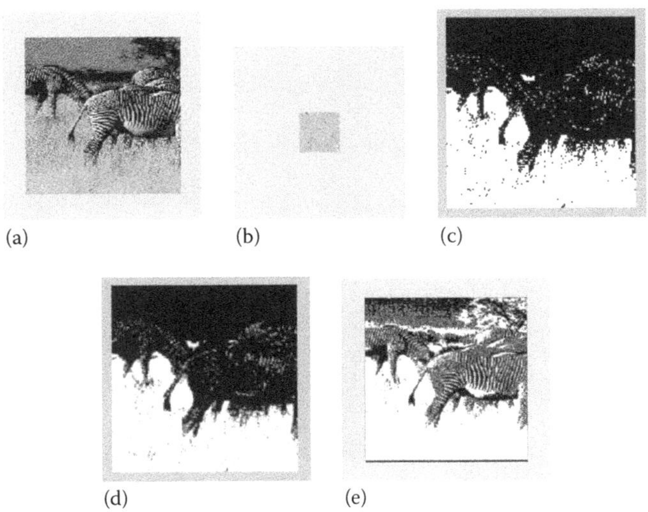

(a) (b) (c) (d) (e)

FIGURE 6.7
(a) **(See color insert following page 144.)** Scene image "zebra"; (b) query image; (c) region-extracted image, the continuous white portion; (d) region-extracted image after using Gaussian filter; and (e) region-extracted image using Smith's method. (From Chaira, T. and Ray, A.K., *Elsevier Pattern Recogn. Lett.*, 12, 1943, 2003. With permission.)

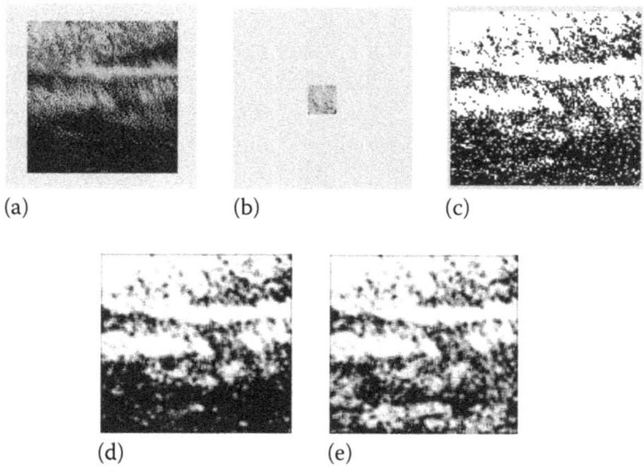

FIGURE 6.8
(a) **(See color insert following page 144.)** Scene image "grassland"; (b) query image; (c) region-extracted image, the continuous white portion; (d) region-extracted image after using Gaussian filter; and (e) region-extracted image using Smith's method. (From Chaira, T. and Ray, A.K., *Elsevier Pattern Recogn. Lett.*, 12, 1943, 2003. With permission.)

Example 6.5

Figure 6.8a is a multicolored scene image of grassland of size 128×128. Grasses are not uniformly distributed and the query image in Figure 6.8b is the grassland

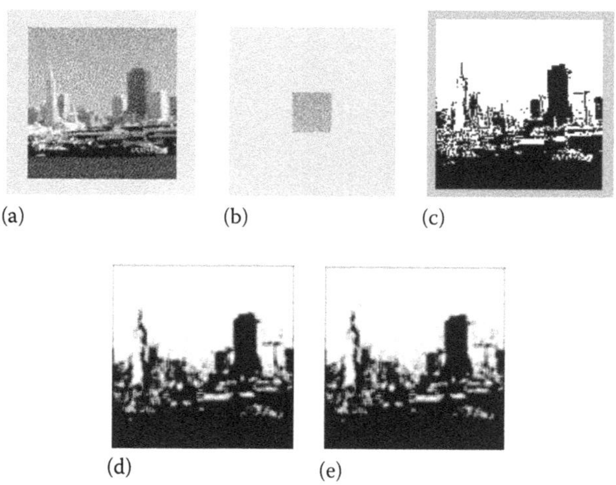

FIGURE 6.9
(a) **(See color insert following page 144.)** Scene image "doc cage city"; (b) query image; (c) region-extracted image, the continuous white portion; (d) region-extracted image after using Gaussian filter; and (e) region-extracted image using Smith's method. (From Chaira, T. and Ray, A.K., *Pattern Recogn. Lett.*, 12, 1943, 2003. With permission.)

region. The region-extracted image in Figure 6.8d is the image that corresponds to the grassland region. This figure is taken from VisTex (Vision Texture) image database. It is seen that the fuzzy method gives better results than the nonfuzzy method.

Example 6.6

Figure 6.9a is a multicolored scene image of doc cage city of size 128×128. The query image is the sky region shown in Figure 6.9b. The region-extracted image in Figure 6.9d is the image that corresponds to the sky region. This figure is taken from VisTex (Vision Texture) image database.

6.4 Summary

Match-based segmentation is used to search a given pattern or model effectively in the scene image. The best match is obtained when the model or the query image is present in the scene image. Match-based segmentation is suitable in an unconstrained image where the image is poorly defined. Back projection algorithm is used for match-based segmentation, which is a class of algorithm designed to detect the regions within scene image with feature histogram that is similar to the histogram of the model image. Nonfuzzy or crisp match-based segmentation methods are proposed by many authors where back projection algorithm is used. In most of the real images, regions are not properly defined and in such cases fuzzy methods may provide better results. Fuzzy match-based segmentation is discussed where back projection algorithm is accompanied with several fuzzy measures. Examples on fuzzy methods are provided on color scene and monochrome images.

References

1. Brodatz, P., Brodatz Textures, http://www.ux.uis.no/~tranden/brodatz.html
2. Chaira, T. and Ray, A.K., Fuzzy approach to color region extraction, *Pattern Recognition Letters*, 12(24), 1943–1950, 2003.
3. Chaira, T. and Ray, A.K., Region extraction using fuzzy similarity measures, *The International Journal of Fuzzy Mathematics*, 11(2), 601–608, 2003.
4. Chaira, T. and Ray, A.K, Region extraction of remotely sensed colored images using fuzzy set theoretic approach, in *International Conference on Communication Devices and Intelligent Systems (CODIS 2004)*, University of Calcutta, Calcutta, India, Jan. 8–10, 2004.

5. Chaira, T., Image segmentation and color retrieval: A fuzzy and intuitionistic fuzzy set theoretic approach, PhD thesis, Indian Institute of Technology, Kharagpur, India, 2003.
6. Enneser, S. and Medioni, G., Finding Waldo or focus of attention using local color information, *IEEE Transactions on Pattern Analysis and Machine Intelligence*, PAMI-17(8), 805, 1995.
7. Smith, J.R., Integrated spatial and feature image systems: Retrieval, analysis and compression, PhD thesis, The Columbia University, New York, 1997.
8. Smith, J.R. and Chang, S.F., Integrated spatial and feature image query, *Multimedia System Journal*, 7(2), 129–140 (1999).
9. Smith, J.R. and Chang, S.F., Tools and techniques of color image retrieval, in *Proceedings of the of SPIE, Storage and Retrieval for Image and Video Databases*, vol. 2670, pp. 1–12, 1996.
10. Swain, M.J. and Ballard, D.H., Color indexing, *International Journal of Computer Vision* 7(1), 11–32 (1991).
11. Tolias, Y.K. et al., Generalized fuzzy indices for similarity matching, *Fuzzy Sets and Systems*, 120, 255–270, 2001.
12. Tversky, A., Features of similarity, *Psychological Review*, 84(4), 327–352, 1977.

7

Fuzzy Edge Detection

7.1 Introduction

Edge-based segmentation is a type of segmentation that is based on the information about the edges in an image. It gives the information about the object boundaries using some edge-detecting operators. These operators find the discontinuities in gray level, color, texture, etc. Edge-detection methods simplify the analysis of the image by reducing the data to be processed. Edge strength is calculated by taking the gradient value of the image when there is a sharp change in the pixel values. Edges may present in many directions and so the important requirements for a good edge detecting operator are that they should be capable of being tuned to act at any desired scale. So the edge detectors are such that they should consider all the edge directions to detect all the edges properly. The most common problem in edge detection is the presence of noise in the image where an edge due to noise will also be detected and these false edges will be misinterpreted as the edges of an image.

There are many approaches to edge detection. It may be done using search-based method where the edge strength is calculated using first-order derivative like gradient magnitude and then searching for local maxima. Some first-order edge-detection operators are Roberts, Sobel, Prewitt, etc. [8,14]. The other prominent methods used for edge detection are the zero crossing of the second-order derivative or Laplacian of the Gaussian, or the difference of Gaussian in the image. Laplacian of Gaussian [14] produces continuous closed boundaries, being the zero set of the implicit function. However, since it is a second derivative operator, it is extremely noise sensitive.

An image may contain noise; thus whether the pixel be an edge pixel or a noise pixel, it depends on the gray value of the pixel and its surrounding pixels. Smoothing is necessary to remove the noise present in the image and Gaussian smoothing is the most common method used for smoothing.

7.2 Methods for Edge Detection

As this chapter is related to fuzzy set, in this section, we will discuss briefly the several ways to detect the edges of an image. Methods may be thresholding based where the edges are detected using a threshold, or Hough transform based where straight lines and shapes are detected or boundary based or fuzzy method based where the edges are considered vague.

7.2.1 Thresholding-Based Methods

After computing the edge strength of an image, the gradient image is thresholded to find the presence of edges. Threshold may vary from image to image since the gray level in the neighborhood of a pixel changes from image to image. It may be selected globally or locally. Local techniques use the operators on local image neighborhood and global techniques use the image as a whole to extract edge information. Both the methods have disadvantages. Global methods give poorer results whereas local methods produce many edges as it picks all the information present in the local area. Most of the edge detectors use threshold.

The purpose of thresholding is to keep the edge pixels whose gradient magnitude is above a threshold T. If the threshold T is too low, many noisy edges will be detected by picking irrelevant features from the image. If T is too high, some important edges will miss out and result in fragmented edges. In both the cases, for T too low or too high, the edge-detected results are undesirable. Good edge-detection methods detect the edges of an image by using smoothing parameter (σ) to remove the noise and thresholding parameter (T) to threshold the gradient image. Depending on the type of image, the thresholding parameter or the scale of smoothing parameter is adjusted to minimize the unwanted noise and to produce well-formed edges. In the case of Gaussian smoothing, if the standard deviation of the Gaussian kernel is large, the noise becomes less sensitive and thus the localization of edges becomes worse. Most of the edge detectors uses threshold. Although there are various methods for edge detection [3,6,7,12,13] but the gradient magnitude is the most commonly used method for thresholding. Canny [3] used another edge-thresholding technique called hysteresis thresholding where two thresholds are used to remain the edges connected as far as possible.

Several researchers proposed different techniques of edge evaluation. Abdou and Pratt [1] suggested a comparison method for various edge-detection methods using Pratt's figure of merit. Pratt's figure of merit is actually used to compare with the other edge-detected image and is defined as

$$F = \left(1/\max(I_A, I_I)\right).\sum_{i=1}^{I_A} 1/\left(1 + kd_i^2\right)$$

where

I_A is the image edge-detected point
I_I is the ideal edge-detected point
$k > 0(1/9)$, d_i is the edge deviation for the ith edge-detected pixel

7.2.2 Boundary Method

Boundary methods mainly follow the gradient features of the spatial positions of an image for segmenting the structures of an image. A global shape measure is used when the whole boundary is considered [4]. The first approach to boundary method for edge detection is the use of parameterized templates called rubber masks. But the most popular method of detecting whole boundary, is the snakes by Kass et al. [9]. Kass developed a novel technique called snake for image segmentation that had excelled more conventional techniques. Kass proposed the concept of a *snake*, which is an active contour model which is "an energy minimizing spline guided by external constraint forces and influenced by image forces that pull it toward features such as lines and edges."

7.2.3 Hough Transform Method

It is another way of finding the boundaries of an image. If an image consists of objects with known shapes and sizes such as squares, circles, rectangles, etc., segmentation procedure may be viewed as a problem of finding this object within an image. One possible way is to move masks of exact shapes and sizes over the image and find the correlation or match between the image and the mask. But this specified mask might differ from the objects present in the image. Hough transform is an effective feature extraction method that can solve the problem. The basic idea of this technique is to find curves like straight lines, polynomials, circles, etc., in a suitable parameter space. It was originally used for parameterizable shapes like curves, but was extended later to include general shapes [2]. Maxima in Hough space signify possible shape features. The advantage of this method is that it is insensitive to gaps and noise and the drawback is the computational requirement that goes up with an increase in the number of parameters.

7.3 Fuzzy Methods

Fuzzy edge detection is another approach to edge detection that considers the image to be fuzzy. In most of the images where the edges are not clearly defined, that is, edges are broken, vague, or blurred etc., and in that case edge detection becomes very difficult. Especially in medical images, where the images are poor contrasted, edge detection becomes very crucial. Because improper selection of edges may lead to wrong diagnosing the diseases. In those cases, fuzzy set theory is very useful that takes into account the vagueness and ambiguity present in the image. This uncertainty is represented in the form of membership function and then the edges are detected. Edges may be detected using fuzzy edge detectors that use some fuzzy templates or fuzzy reasoning that uses some linguistic variables. There are various fuzzy edge–detection methods in the literature. Here we shall discuss several fuzzy edge–detection methods.

7.3.1 Fuzzy Sobel Edge Detector

As in the crisp Sobel detector, there is Fuzzy Sobel edge detector suggested by Khamy et al. [12]. Sobel method uses two 3×3 masks convolved with the image. Two masks calculate the gradient in two directions and the final gradient is calculated by combining the two results. Edges are detected using a threshold that depends on trial and error method by the user. In fuzzy Sobel method, the image is divided into two regions. If the pixels are having high difference in the gray level with their neighborhood region, then the pixels are classified in a fuzzy edge region and if the pixels are less difference of gray level with their neighborhood region, then the pixels are classified in a fuzzy smooth region. The boundaries of the two regions are determined from the four threshold values that are derived from difference histogram. The difference histogram is calculated by calculating the maximum difference in the gray level value of each pixel in all the eight directions which is as follows:

$$\text{diff}(x,y) = \max(I(x,y) - I(x+i,y+j))$$

where

- $\text{diff}(x, y)$ is the maximum difference in the pixel with its eight neighborhood pixels and $i, j = -1, 0, 1$
- $I(x, y)$ is the gray level of the image pixel at location (x, y) and $I(x + i, y + j)$ is the gray level of the image pixel in the eight directions

The four threshold values such as low threshold (LT), high limit (HL), low limit (LL), and high threshold (HT) are calculated from the difference histogram. The values of the difference histogram normally do not range from 0 to 255. The range in the difference histogram, where there is a 20% decay of the peak value, is marked and the lower value of the range is assigned as the LT region and the higher value of the range is assigned as the high threshold region. As the spacings between the thresholds are considered equal, high limit and low limit are obtained accordingly. With these four threshold values, two fuzzy regions—fuzzy smooth region and fuzzy edge region are constructed. The gradient of the input image, i.e., $G(x, y)$, is initially obtained using Sobel operator. The output of the fuzzy modified edge detector is derived using fuzzy reasoning. The fuzzy rules are as follows:

$$\begin{aligned} R(x,y) &= 255, \quad \text{if } G(x,y) \geq \text{HT} \\ &= 0, \quad \text{if } G(x,y) \leq \text{LT} \\ &= G(x,y) \cdot \max(\mu_{\text{SFR}}(x,y), \mu_{\text{EFR}}(x,y)) \quad \text{otherwise} \end{aligned}$$

where

μ_{SFR} and μ_{EFR} are the membership functions of the image smooth and edge regions as shown in Figure 7.1

$G(x, y)$ is the gradient value using Sobel operator

$R(x, y)$ is the resultant pixel at location (x, y)

Example 7.1

Two examples on Fuzzy Sobel edge detector are shown in Figures 7.2 and 7.3 that will show the efficacy of Sobel edge detector. Figure 7.2a is the human lung image of size 90×90 where the two lungs are present. Figure 7.2b is the fuzzy edge–detected image. Figure 7.3a is the human blood cell image of size 90×90 where few blood cells are seen. Figure 7.3b is the fuzzy edge–detected image.

7.3.2 Entropy-Based Fuzzy Edge Detection

The method, suggested by Khamy et al. [11] is a two level thresholding problem where the gradient threshold is used to partition the image in two regions i.e., smooth region and edge regions. The threshold is selected using the principle of minimum entropy. Their method is an extension of [15] three level image thresholding using fuzzy 3 partition, probability partition, and entropy. Initially a gradient image $G(x, y)$ is formed from the image $I(x, y)$ of size $M \times N$ using a Sobel operator. Then the normalized gradient image is calculated to restrict the values between 0 and 100 as follows:

$$G_{\text{norm}}(x, y) = \text{round}((G(x, y)/\max(G)) \times 100)$$

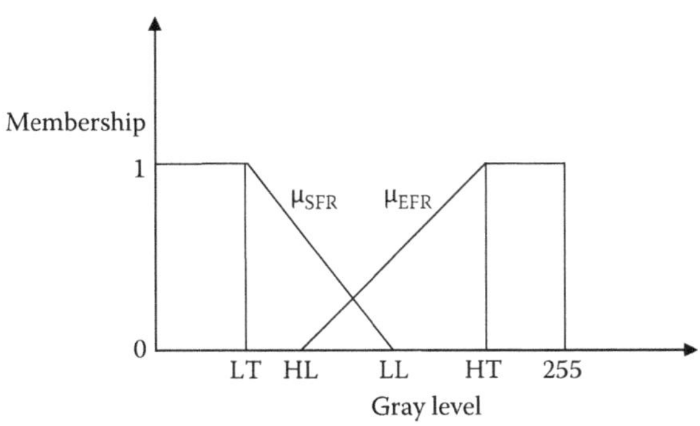

FIGURE 7.1
An example of image regions. (From Khamy, E.L. et al., Modified Sobel fuzzy edge detector, in *Proceedings of IEEE MELCON, 17th National Radio Science Conference*, C32-1-9, Egypt, 2000. With permission.)

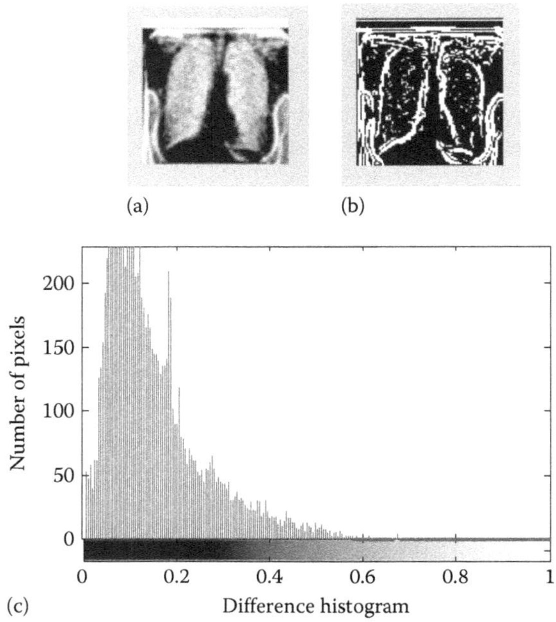

FIGURE 7.2
(a) Lung image, (b) edge image, and (c) difference histogram.

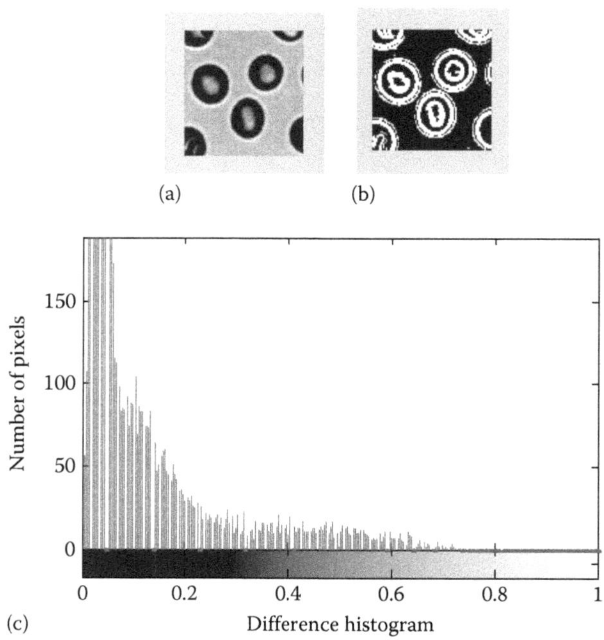

FIGURE 7.3
(a) Blood cell image, (b) edge image, and (c) difference histogram.

where

$$x = 0,1,2,...,M-1, \quad y = 0,1,2,...,N-1$$

G_{norm} is defined by its domain D and range R, such that $G_{norm}(x, y) \in R$, $\forall (x, y) \in D$, where $D = \{D(x, y), x = 0, 1, 2, \ldots, M-1, y = 0, 1, 2, \ldots, N-1\}$ and $R = \{0, 1, 2, \ldots, K\}$, $K = 100$. The histogram of G_{norm} is defined by $HG_{norm} = \{hg_0, hg_1, \ldots, hg_K)$ where $hg_k = n_k/MN$, $k = 0, 1, 2, \ldots, K$, and n_k is the frequency of occurrences of the value k in G_{norm}. The probability partition (PP) of the normalized gradient image is defined as $PP_g = \{D_0, D_1, \ldots, D_K)$ which is characterized by the probabilistic distribution

$$p_k = P(D_k) = hg_k, \quad k = 0,1,2,..., K$$

where $D_k = \{(x, y)\text{: } G_{norm}(x, y) = k, (x, y) \in D\}$

In this method, the image is partitioned into two regions-D_{SMOOTH} and D_{EDGE} for low gradient and high gradient values respectively. It is required to find the probabilistic fuzzy 2-partition that partitions the domain D in two regions with probabilistic partition, $PP_2 = \{D_{SMOOTH}, D_{EDGE}\}$. The partition's probability distribution for these two regions are p_{SMOOTH} i.e., $p_{SMOOTH} = P(D_{SMOOTH})$ and p_{EDGE} i.e., $p_{EDGE} = P(D_{EDGE})$. The two fuzzy partitions are characterized by two trapezoidal membership functions, μ_{SMOOTH} and μ_{EDGE}, with parameters t_1 and t_2 such that $0 \le t_1 \le t_2 \le 100$ as shown in Figure 7.4. The membership function represents the conditional probability that a pixel is classified in two regions with the condition that it has a gradient k, that is, $\mu_{SMOOTH}(k) = p_{SMOOTH|k}$ and $\mu_{EDGE}(k) = p_{EDGE|k}$. If t_1 and t_2 are selected, the fuzzy 2 partition, $PP_2 = \{D_{SMOOTH}, D_{EDGE}\}$ is found and probabilistic distribution of the two partitions are $p_{SMOOTH} = \sum_k hg_k \cdot \mu_{SMOOTH}(k)$ and $p_{EDGE} = \sum_k hg_k \cdot \mu_{EDGE}(k)$.

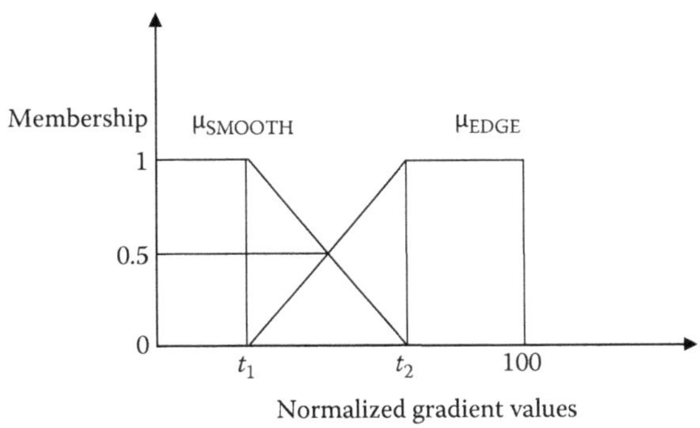

FIGURE 7.4
Membership function for smooth and edge regions. (From Khamy, E.L. et al., Fuzzy edge detection using minimum entropy, in *Proceedings of 11th Mediterranean Electrotechnical Conference MELECON*, Cairo, Egypt, 2002. With permission.)

Next, the minimum entropy criterion is used to select the threshold since edge representation is the compact representation of images. The entropy is defined as

$$H(t_1,t_2) = -p_{\text{SMOOTH}} \cdot \log_2(p_{\text{SMOOTH}}) - p_{\text{EDGE}} \cdot \log_2(p_{\text{EDGE}}).$$

The parameter values corresponding to the minimum entropy is selected. The condition for the entropy $H(t_1,t_2)$ to reach minimum or maximum is

$$\partial H(t_1,t_2) \Big/ \partial(t_1) = 0 = -\left(\partial p_{\text{SMOOTH}} \Big/ \partial t_1 \times \log\left(p_{\text{SMOOTH}}/1 - p_{\text{SMOOTH}}\right)\right)$$

$$\partial H(t_1,t_2) \Big/ \partial(t_2) = 0 = -\left(\partial p_{\text{SMOOTH}} \Big/ \partial t_2 \times \log\left(p_{\text{SMOOTH}}/1 - p_{\text{SMOOTH}}\right)\right)$$

The entropy will be minimum when $p_{\text{SMOOTH}} = p_{\text{EDGE}} = 1/2$. When the searching process is complete, there is R set of parameter pairs $(t_1(r), t_2(r))$, $r = 1, 2, 3, \ldots, R$ that satisfy $|p_{\text{SMOOTH}} - 1/2| \le \varepsilon$ and $|p_{\text{EDGE}} - 1/2| \le \varepsilon$, where ε is a small positive integer of the order 0.05. The best set of parameters $(\tilde{t}_1, \tilde{t}_2)$ is one that satisfies $H(\tilde{t}_1, \tilde{t}_2) = \min_{r=0,1,2,\ldots,R} \left(H(t_1(r), t_2(r))\right)$.

The edge image is calculated as

$$\begin{aligned} \text{Edge\_image}(x,y) &= 1 \quad \text{if } G_{\text{norm}}(x,y) \ge t \\ &= 0 \quad \text{if } G_{\text{norm}}(x,y) < t \end{aligned}$$

where $t = \left(\tilde{t}_1 + \tilde{t}_2/2\right)$ but the drawback is that unwanted edges are detected. These may be removed by sliding a 5×5 window over the entire edge-detected image. Within each window position at point (x, y) the number of edge points is calculated and then at each window location, a rule is followed.

If the number of edge points $\ge S$, edge points are not considered (normally $S = 10$ is appropriate for 5×5 window); else remove these points that represent the unwanted edges.

7.3.3 Fuzzy Template Based Edge Detector

Two types of fuzzy edge detectors are described.

i. Kenneth et al. [10] presented a fuzzy edge detector, FEDGE. Fuzzy edge detector may be viewed as a fuzzified template convolution where a set of fuzzy edge profiles or fuzzy edge templates is designed and these are convolved with the image. The dimension of the templates is less than that of the image.

Initially the intensity of each pixel is normalized so that the values lie in the interval between 0 and 1. The number of fuzzy templates is considered as "n" and these templates denote the possible edge directions of the image. Each template is placed on the image at point (x, y) and the sub-image is the area on the image where the template is placed. Sub-image S or the image window is of the same size as that of the template. To find the existence of an edge at point (x, y) of image I, the surrounding features that are the pixels are compared against the template. The fuzzy template is shifted in position to compute the similarity between the image and the template. Then fuzzy similarity measure between each pixel in the sub-image S and the pixel value of each template T is computed that finds the existence of an edge at (x, y) within the input image I.

The fuzzy templates are as follows:

$$\begin{bmatrix} 0 & 0 & 0 & 0 & 0 \\ 0.5 & 0.5 & 0.5 & 0.5 & 0.5 \\ 1 & 1 & 1 & 1 & 1 \\ 0.5 & 0.5 & 0.5 & 0.5 & 0.5 \\ 0 & 0 & 0 & 0 & 0 \end{bmatrix} \begin{bmatrix} 0 & 0.5 & 1 & 0.5 & 0 \\ 0 & 0.5 & 1 & 0.5 & 0 \\ 0 & 0.5 & 1 & 0.5 & 0 \\ 0 & 0.5 & 1 & 0.5 & 0 \\ 0 & 0.5 & 1 & 0.5 & 0 \end{bmatrix}$$

$$\begin{bmatrix} 0 & 0 & 0 & 0.5 & 1 \\ 0 & 0 & 0.5 & 1 & 0.5 \\ 0 & 0.5 & 1 & 0.5 & 0 \\ 0.5 & 1 & 0.5 & 0 & 0 \\ 1 & 0.5 & 0 & 0 & 0 \end{bmatrix} \begin{bmatrix} 1 & 0.5 & 0 & 0 & 0 \\ 0.5 & 1 & 0.5 & 0 & 0 \\ 0 & 0.5 & 1 & 0.5 & 0 \\ 0 & 0 & 0.5 & 1 & 0.5 \\ 0 & 0 & 0 & 0.5 & 1 \end{bmatrix}$$

Let the pixel at point (x, y) in the template T be (x_T, y_T) and the corresponding point in the image be (x_I, y_I). Then the similarity measure between each pixel in the template and the image window is calculated as follows

$$H_T(x,y) = 1 - \left|\text{pixel}_I(x_I, y_I) - \text{pixel}_T(x_T, y_T)\right|,$$

where $\text{pixel}_I(x_I, y_I)$ and $\text{pixel}_T(x_T, y_T)$ are the pixels at position (x, y) for the image and the template.

This is calculated for all the templates. The similarity measures for all the templates are combined together by using a simple max operator for finding the existence of edges.

$$H(x,y) = \max_{T=1}^{T=n}\left(H_T(x,y)\right), \quad T = 0,1,2,\ldots n, \quad n = \text{number of templates.}$$

The new image so formed is alpha_cut or threshold to represent the high support for the existence of the edges i.e., those values which are below the alpha_cut are set as 0 and those are above the alpha_cut are set as 1.

Example 7.2

An example on Fuzzy edge detector is shown in Figure 7.5a to illustrate the fuzzy edge detector. Figure 7.5a is a rice image of size 128×128. Figure 7.5b is the fuzzy edge–detected image where the boundaries of the rice seeds are extracted.

ii. Another fuzzy template based edge–detection method is suggested by Chaira and Ray [5–7] that uses fuzzy divergence between the image window and each of a set of 16 fuzzy templates.

A set of 16 fuzzy templates, each of size $q \times q$ (here $q=3$), representing the edge profiles of different types are used as shown in Figure 7.6.

The choice of templates is crucial which reflects the type and direction of edges that occur. The templates are the examples of the edges, which are also the images. a, b, and 0 represent the pixels of the edge templates. The values of $a=0.3$, $b=0.8$ are chosen (arbitrarily) to ensure good edge detected image. The size of the templates is less than

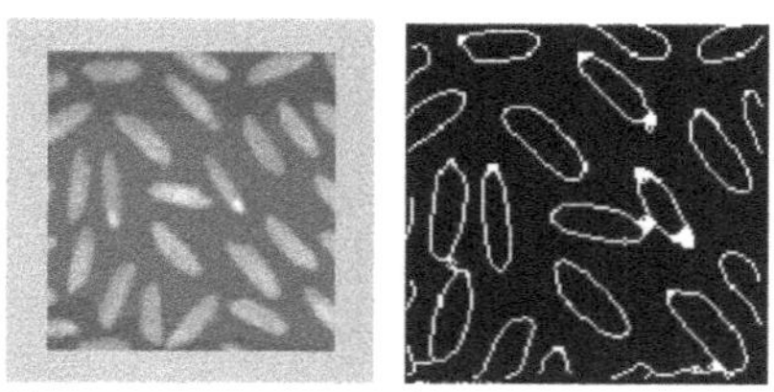

(a) (b)

FIGURE 7.5
(a) Original image, (b) edge-detected rice image.

$$\begin{bmatrix} 0 & b & a \\ 0 & b & a \\ 0 & b & a \end{bmatrix} \begin{bmatrix} a & a & a \\ 0 & 0 & 0 \\ b & b & b \end{bmatrix} \begin{bmatrix} a & a & b \\ a & b & 0 \\ b & 0 & 0 \end{bmatrix} \begin{bmatrix} b & b & b \\ 0 & 0 & 0 \\ a & a & a \end{bmatrix} \begin{bmatrix} b & a & a \\ 0 & b & a \\ 0 & 0 & b \end{bmatrix} \begin{bmatrix} b & a & 0 \\ b & a & 0 \\ b & a & 0 \end{bmatrix} \begin{bmatrix} a & 0 & b \\ a & 0 & b \\ a & 0 & b \end{bmatrix} \begin{bmatrix} 0 & 0 & 0 \\ b & b & b \\ a & a & a \end{bmatrix}$$

$$\begin{bmatrix} a & a & a \\ b & b & b \\ 0 & 0 & 0 \end{bmatrix} \begin{bmatrix} a & b & 0 \\ a & b & 0 \\ a & b & 0 \end{bmatrix} \begin{bmatrix} 0 & 0 & 0 \\ a & a & a \\ b & b & b \end{bmatrix} \begin{bmatrix} 0 & a & b \\ 0 & a & b \\ 0 & a & b \end{bmatrix} \begin{bmatrix} b & b & b \\ a & a & a \\ 0 & 0 & 0 \end{bmatrix} \begin{bmatrix} b & 0 & a \\ b & 0 & a \\ b & 0 & a \end{bmatrix} \begin{bmatrix} b & 0 & 0 \\ a & b & 0 \\ a & a & b \end{bmatrix} \begin{bmatrix} 0 & 0 & b \\ 0 & b & a \\ b & a & a \end{bmatrix}$$

FIGURE 7.6
Set of 3×3 templates.

the size of the image. It is observed that, on increasing the number of templates, there is no remarkable change in the edge results. In the method, initially the image is normalized. The center of each template is placed at each pixel position (i, j) over the normalized image. The fuzzy divergence measure Div_measure(i, j) at each pixel position (i, j) in the image, where the template was centered, is calculated between each of the elements of the image window (same size as that of the template) and the template and the minimum value is selected. This is repeated for all the 16 fuzzy templates and the maximum value among the 16 divergence values is selected. This may be represented mathematically as

$$\text{Div\_measure}(i,j) = \max_{N}\left[\min_{r}(\text{Div}(A,B))\right] \tag{7.1}$$

where A represents the chosen window in the test image which is of same size as that of the template and B is the template. The fuzzy divergence between image window A and template B, Div(A, B), is calculated by computing the divergence between each of the elements a_{ij} and b_{ij} of image A and of template B, respectively:

$$\text{Div}(a_{ij}, b_{ij}) = \begin{bmatrix} 2-\left(1-\mu_A(a_{ij})+\mu_B(b_{ij})\right)\cdot e^{\mu_A(a_{ij})-\mu_B(b_{ij})} \\ -\left(1-\mu_B(b_{ij})+\mu_A(a_{ij})\right)\cdot e^{\mu_B(b_{ij})-\mu_A(a_{ij})} \end{bmatrix} \tag{7.2}$$

N=number of templates, r=elements in the square template, i.e., $3^2=9$. However, the size of the square template may vary.

Membership values of the (i, j)th pixel of the normalized image A is represented as $\mu_A(a_{ij})$ and that of the values of the template image B is $\mu_B(b_{ij})$. Max–min value from Equation 7.1 is calculated for all pixel positions. A fuzzy divergence matrix is formed. The divergence matrix is thresholded and thinned to obtain an edge-detected image.

The edge detection procedure is summarized in Algorithm 7.1.

Algorithm 7.1 Fuzzy edge detection [7]

Step 1. Apply the edge templates over the image by placing the center of each template at each point (i, j) over the normalized image.

Step 2. Calculate the fuzzy divergence value between each of the elements of the image window and the template (same size as that of template) and choose the minimum value.

Step 3. Repeat step 2 for all the 16 fuzzy templates.
Step 4. Choose the maximum among all the 16 (total number of templates) minimum divergence values.
Step 5. Position the maximum value at the point where the template was centered over the image.
Step 6. Repeat steps 2–5 for all the pixel positions (considering the border pixels by taking the mirror values of the image) to form a new divergence matrix.
Step 7. Threshold the divergence matrix and thin.

Edge detection is very useful in biomedical imaging that helps the physicians to diagnose different types of diseases. Suppose, in the ultrasound images of tumor or cyst or calcification, edge detection will clearly detect the tumor or cyst or other regions. In chronic obstructive pulmonary disease (COPD), edge detection will show the reduction in the air passage and other abnormalities. Some examples are shown on medical image and real images to show the efficacy of the fuzzy methods.

Example 7.3

Three results on real-time images such as human face image, lung image, and rice image will illustrate the effect of fuzzy method in edge detection [5]. The results on different types of images are shown in Figures 7.7 through 7.9.

Figure 7.7a depicts a "Rice" image of size 128×128. The divergence matrix thresholded at 0.38 is thinned and the result is shown in Figure 7.7b. It is observed that the edges are quite prominent and clear.

The human "lung" image of size 189×189 is shown in Figure 7.8a where the two lungs are visible. The divergence matrix is thresholded at 0.2 and thinned shown in Figure 7.8b. The boundary of the two lungs is clearly detected in Figure 7.8b.

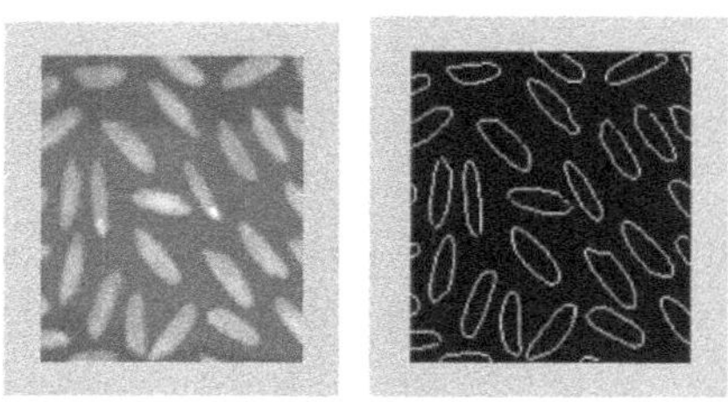

(a) (b)

FIGURE 7.7
(a) Input image, rice and (b) edge-detected image.

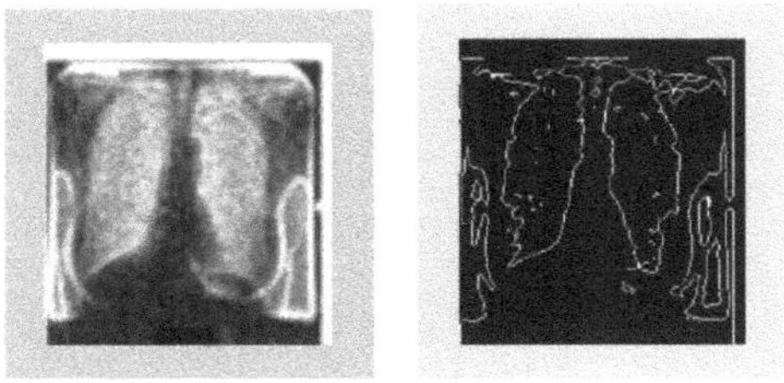

(a) (b)

FIGURE 7.8
(a) Input image, lung and (b) edge-detected image.

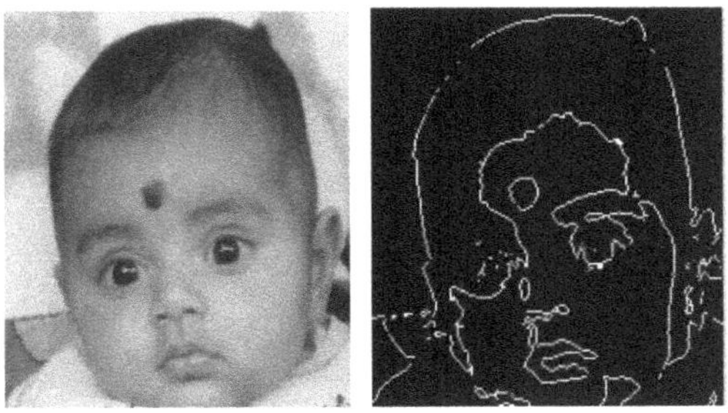

(a) (b)

FIGURE 7.9
(a) Input image, Shruti and (b) edge-detected image.

Baby image "Shruti" of size 256×256 is shown in Figure 7.9a. The divergence matrix is thresholded at 0.16 and thinned shown in Figure 7.9b.

Example 7.4

A result of edge detection on color image "Lena" is shown in Figure 7.10. Edge detection is done for each color channel (R, G, B) and finally they are combined to detect all the edges. The image is thresholded separately for different channels, for example, for red channel $R=0.2$, for green channel $G=0.16$, and for blue channel $B=0.16$. The threshold results for each color channels (R, G, B) are shown and the combinations of the three colors are also shown.

From Figure 7.10, it can be seen that the edge-detected image of the red part gives good information than the green and blue part. In the combined edge-detected image, many edges are seen due to the different color shades present in the image.

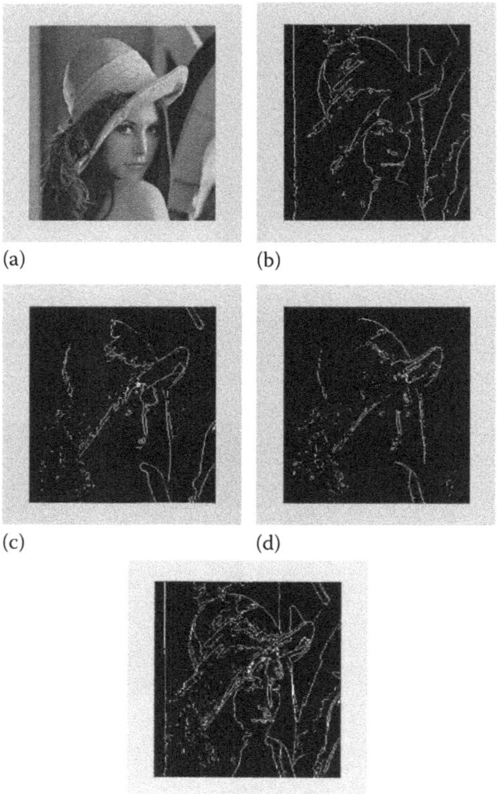

(a) (b) (c) (d) (e)

FIGURE 7.10
(a) **(See color insert following page 144.)** "Lena" image, (b) edge detection for the red channel, (c) edge detection for the green channel, (d) edge detection for the blue portion of the image, and (e) combined edge-detected image.

7.4 Summary

Edge detection is another type of segmentation where the gradient pixels are detected. It finds the boundaries in the image. There are various types of edge segmentation methods—Hough transform, active contour model, thresholding-based methods, and edge detectors. The most common problem in edge detection is that sometimes the edges are not prominent and are vague in nature. As the edges are vague, fuzzy methods are very useful in dealing with this type of problem. There are various fuzzy edge–detection methods such as fuzzy reasoning, fuzzy template based edge detector, fuzzy Sobel edge detector, fuzzy entropy based edge detector. Fuzzy reasoning method uses some fuzzy rules that have the ability to understand natural language. Fuzzy template based edge detector uses a set of fuzzy edge templates that signifies the possible edge directions of an image and uses fuzzy similarity measure to find the edges in an image that corresponds to the templates. Edges are detected depending on the type of fuzzy edge templates. Fuzzy edge detectors are applicable to both the monochrome and color images. Few results are also displayed for both color and black- and-white images.

References

1. Abdou, I.E. and Pratt, W.K., Quantization design and evaluation of enhancement/thresholding edge detectors, *Proceedings of the IEEE*, 67(5), 753–763, 1979.
2. Ballard, D.H. and Brown, C.M., *Computer Vision*, Prentice-Hall, Englewood Cliffs, NJ, 1982.
3. Canny, J., Computational approach to edge detection, *IEEE Transaction on PAMI*, 8(6), 1986, 679–698.
4. Chakraborty, A., Feature and module integration for image segmentation, PhD thesis, Yale University, New Haven, CT, 1996.
5. Chaira, T., Image segmentation and color retrieval: A fuzzy and intuitionistic fuzzy set theoretic approach, PhD thesis, IIT Kharagpur, India, 2003.
6. Chaira, T., De, S., and Salvetti, O., A new measure using intuitionistic fuzzy set using Hausdorff metric and its application to edge detection, *The International Journal of Fuzzy Mathematics*, 15(3), 645–654, 2007.
7. Chaira, T. and Ray, A.K., A new measure using intuitionistic fuzzy set theory and its application to edge detection, *Applied Soft Computing*, 8(2), 919–927, 2008.
8. Heath, M.D., Sarkar, S., Sanokhi, T., and Bowyer, K., Comparison of edge detectors: A methodology and initial study, in *Proceedings of the CVPR*, Tampa, FL, pp. 143–148, 1996.
9. Kass, M., Wilkin, A., and Terzopaulus, D., Snakes: Active contour models, *International Journal of Computer Vision*, 1, 312–331, 1988.
10. Ho Kenneth, H.L. and Ohnishi, N., FEDGE—fuzzy edge detection by fuzzy categorization and classification of edges, in *Fuzzy Logic in Artificial Intelligence, JCAI'95 Workshop*, Selected Papers, pp. 182–196, 1995.
11. Khamy, E.L. et al., Fuzzy edge detection using minimum entropy, in *Proceedings of 11th Mediterranean Electrotechnical Conference MELECON*, Cairo, Egypt, 2002.
12. Khamy, E.L. et al., Modified Sobel fuzzy edge detector, in *Proceedings of 17th National Radio Science Conference* (NRSC 2000), C32-1-9, Minufia, Egypt, 2000.
13. Rosin, P.L., Edge: Saliency measures and automatic thresholding, *Machine Vision and Application*, 9, 139–159, 1997.
14. Sonka, M. et al., *Image Processing, Analysis and Machine Vision*, Brooks/Cole Publisher, London, 2001.
15. Zhao, M. et al., A technique for three level thresholding based on probability partition and Fuzzy 3 partition, *IEEE Transactions on Fuzzy Systems*, 9(3), 469–479, 2001.

8

Fuzzy Content-Based Image Retrieval

8.1 Introduction

The recent development of multimedia technologies such as medical diagnosis, imaging techniques, remote sensing, weather forecasting, entertainment, and other services has gained popularity. We can see a vast type of images on the Internet: satellite photos used in weather forecasting, different logos of different companies, flags of different companies, and so on. These image collections form different types of database. During the early days of computing, this advancement was similar to the increase in the alphanumeric data that led to the development of a Database Management systems. Traditional Database Management Systems were designed to organize alphanumeric data into interrelated collections so that information retrieval and storage could be done efficiently. This technology, however, is not well suited to the management of multimedia information, because of the following reasons:

1. There exists a large diversity of data types and formats associated with images, video, and other multimedia data.
2. Size of static and dynamic images is often quite large.
3. There are lot of difficulties in the automatic extraction of the semantic information from image and video data.

Since 1970, content-based retrieval is an active research in the field of image and video processing. During the late 1970s, traditional image database is formed by the annotated keywords of the image and stored in retrieval keys. The approach to searching and retrieval of a text-annotated image database manually is slow, laborious, and expensive. In addition, textual annotations cannot encode all the information available in the image and also the content of the image is not properly captured using a small number of keywords. To use this widely used multimedia information, a system is required for retrieving and indexing images, because conventional database cannot provide efficient performance. The problem assumes adequate significance because the automatic segmentation of an image into a set of homogenous regions representing objects and then retrieving the regions is a difficult problem in computer vision. An overview on content based retrieval is found in [9].

In the 1990s, content-based retrieval system (CBIR) was proposed to overcome the difficulties in conventional system. Ideally CBIR, should automatically extract the semantic information of the image to meet user's requirement. But, indexing and retrieval of images is a difficult problem. As for example, recognition and classification of a flower or zebras picked from a collection of pictures are still a difficult problem in image understanding. Thus content-based retrieval is performed using low-level image features such as color, texture, and shape that substitute the high-level, text-based, manual annotations.

The task of image retrieval using similarity is the mutual correspondence between two images in a set of database images. Content-based query systems process a query image and the images in the database and search for the query image, which resembles the image in the database. The performance of the CBIR system depends on several factors:

1. Extracting a set of features that constitute the image
2. Identifying suitable similarity measure based on image distance function

8.2 Color Spaces

Color image constitutes a large number of colors as compared to gray images. Color is a perceptual phenomenon related to human response to different wavelengths. A vast amount of information is embedded in a color image. For analyzing color images, it is important to map the colors in an appropriate color space, so that each point in the color space represents only a specific color. There are different types of color spaces. The choice of a color space by which the images will be represented, analyzed, and compared, depends on the type of user-defined problem. Color spaces are classified in three main categories: (1) hardware-oriented, (2) user-oriented, and (3) uniform color spaces.

Hardware-oriented models are defined on the basis of the properties of the devices used to reproduce colors in computer screen, TV, monitor, etc. The most common color space in a hardware-oriented model is RGB (red, green, blue) [2,3] model. It is used in the output of most of the cameras. RGB space is device dependent, that is, it displays the colors, which are not only dependent on RGB color but also on the device specification and light intensity. Figure 8.1a shows the RGB color model in which the main diagonal represents the gray values from black to white and the axes represent varying brightness of R,G,B. Any color point inside the cube is represented by the weighted sum of red, green, and blue.

User-oriented color spaces [3,10] are based on human color perception. Human visual system is sensitive to the range of electromagnetic radiation.

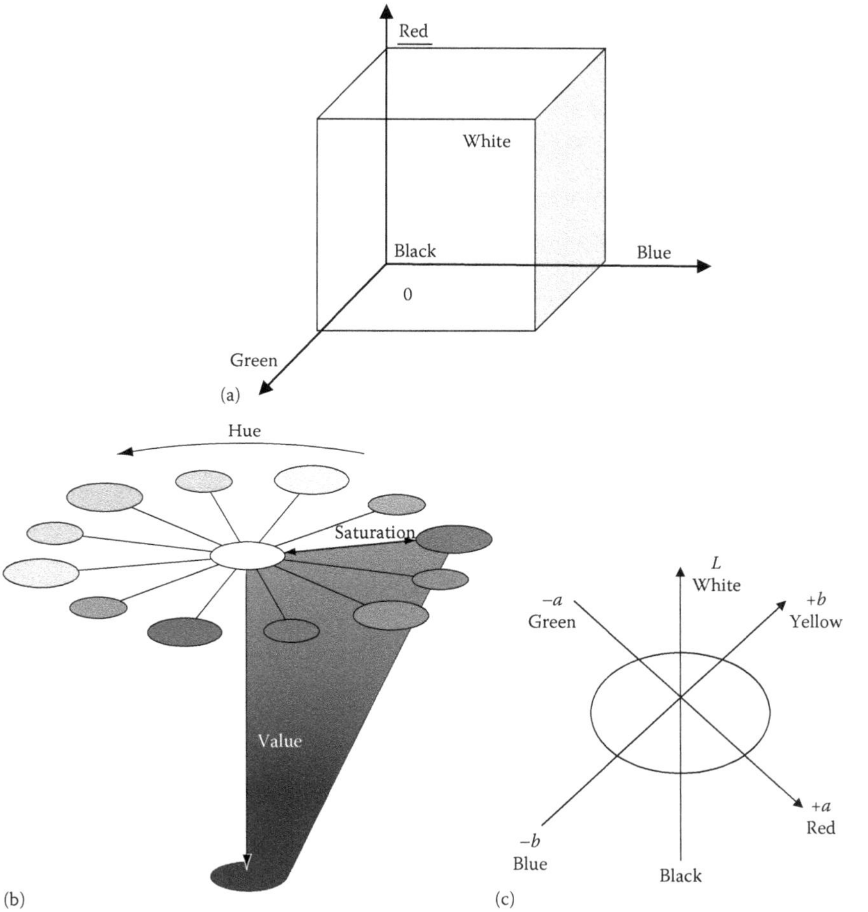

FIGURE 8.1
(a) RGB model, (b) **(See color insert following page 144.)** HSV model (http://www.ncsu.edu.scivis/lessons/colormodels/color_modelse.html. With permission.), and (c) CIELAB color model.

They exploit those points that are used by humans to distinguish one color from another. These are

1. Hue: It is related to the wavelength of the light that produces the visual sensation of red, yellow, green, and blue, or a combination of two of them. These colors are represented in angular form i.e., from 0 degree to 360 degrees.
2. Saturation: It denotes the purity of the color which is related to the standard deviation around the dominant wavelength. It indicates the amount of gray present in the color. If the color is bright, then the saturation is high implies that the color contains less gray and vice versa. Its value ranges from 0(gray) to 1(primary color).

3. Value: It denotes the brightness of the color, that is, how light or dark the color is and ranges from 0 (dark) to 1 (bright). It is related to the amount of white in the color.

Some of the user-oriented spaces are the hue, saturation, intensity (HSI) and hue, saturation, value (HSV) color spaces.

Figure 8.1b shows a 3-D geometrical representation of the HSV model. Hue is represented by angular position from 0° to 259°. Red starts at 0°, green starts at 120°, and blue starts at 240°. Complementary colors such as yellow, cyan, and magenta are in between. On the outer edge of the hue wheel, hues are pure. As we move toward the center of the wheel, the hue becomes less and less and at the center, there is no hue. The saturation is determined by the distance from the center (fully de-saturated) to the hue wheel (fully saturated). Value is defined as the distance along the gray line, to know how light or dark the color is.

Uniform color spaces are related to the human perception. It considers the color difference that is consistent with the differences perceived by humans. These are $L^*a^*b^*$ and $L^*u^*v^*$ color spaces of CIE [2,6]. CIE $L^*a^*b^*$ (CIELAB) is the most complete color model used to describe all the colors visible to the human eye. This color space is not dependent on a particular device. The L^* axis is known as the lightness that extends from 0 (black) to 100 (white). The other two coordinates a^* and b^* do not have numerical values. Positive a^* is red and negative a^* is green while positive b^* is yellow and negative b^* is blue. These sensations are used to develop a complementary color system. It is based on the differences of three elementary color pairs: red–green, yellow–blue, and black–white.

CIELAB is shown in Figure 8.1c. The a axis of the CIELAB color space extends from green ($-a$) to red ($+a$) and the b axis from blue ($-b$) to yellow ($+b$). The brightness, L, increases from bottom (black) to the top (white) of the –3-D model.

8.3 Content-Based Color Image Retrieval

There are many ways one can retrieve the images depending on the user's need. The methods are classified into three main groups: (1) global-based approaches, (2) partition-based approaches, and (3) regional-based approaches. This classification is based on the representation of color information present in an image.

8.3.1 Global-Based Approach

Global approach is the most efficient one in terms of visual features extraction, space overload, and comparison of images. It takes into account the global histogram of the entire image. The global color histogram is the most simple and

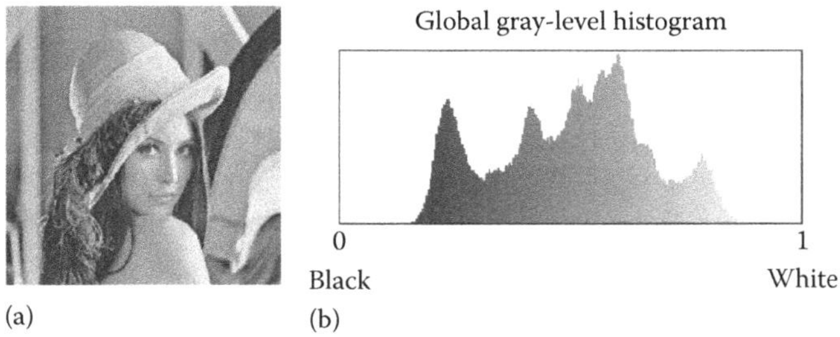

FIGURE 8.2
(a) 'Lena' image and (b) its global histogram.

well-known approach to color retrieval that describes the color distribution of an image as shown in Figure 8.2. The figure shows a 'Lena' image and its equivalent histogram in gray scale. Similar to this histogram, color histogram is constructed separately for each color channel i.e., R, G, B and it represents the color distribution of each channel in the image. Histogram is a k-dimensional vector, where k is the number of colors represented by the histogram.

Histograms of two images are correlated by computing the color similarity using some distance function. If the distance between two image histograms is less, the images are similar and if the distance is more, the images are dissimilar. Histogram formation may be of two types: global and local histogram. Global color histogram finds the color distribution across the entire image using some similarity function and thus it is compact and robust. Local histogram computes the color distribution locally in the partitions or blocks of the image. These two techniques are suitable for different types of image queries. If the retrieval type considers the overall colors in the image regardless of the spatial distribution of colors in the image, global color distribution is useful otherwise local distribution is useful.

The main advantage of this approach is that it is very efficient in terms of visual feature extraction, comparison, and is also invariant to image rotation and translation. The histogram-based technique provides fast and better results, but the disadvantage is that it does not consider the spatial information of color. There are lot many retrieval methods in literature using both fuzzy and crisp methods. Vertan and Boujemma [12] used fuzzy color histograms for image retrieval by accounting the uncertainty of the color components. Images are compared by means of distances between fuzzy images using fuzzy distance function. Tolias [11] used fuzzy similarity metric for image retrieval that uses both color and shape on the HSV model.

8.3.2 Partition-Based Approach

Partition-based approach is advantageous to global approaches. It divides the image into different windows where the spatial information of color is located properly. The image is divided into several nonoverlapping equal sized rectangular blocks or cells of sizes $m \times n$. The dominant colors are extracted for each block. The color distribution of each block is described with the local color histogram by calculating the distance between the local color histogram of equivalent cells of the two images which are at same spatial position. This representation of the color information has the advantage of spatially locating the colors inside the image. Two images are similar if they have similar representative colors in the corresponding blocks. So, comparison is performed only when the corresponding blocks in the two images have a representative color and the extent of similarity is computed by finding the distance between the two images. In fuzzy set theory, the image is divided into blocks and each block is characterized by fuzzy features that can be colors, shapes, or sizes. The fuzzy features are defined for both the images and the comparison is done by finding the similarity between the equivalent blocks of the two images that are at same spatial location.

8.3.3 Regional-Based Approach

Regional color-based image retrieval systems segment the image into regions and retrieve images based on the similarity between the regions. The most common method is to compare the features that can be colors, shape, or size of the regions. This method focuses too much on region-based similarity. A fuzzy region–based feature matching segments an image into regions and each region is characterized by a set of fuzzy features that reflect color, texture, and shape properties. Thus each image is characterized by a class of fuzzy feature that characterizes the gradual transition between the regions. Membership degrees are computed for each feature in the feature space. The matching process consists of three steps. First, each fuzzy feature of the query image is then compared with all the features of the target image. Likewise, each fuzzy feature of the target image is compared with all the features of the query image. The similarity between two images is the weighted sum of the distances (similarity degree) for both the steps [5]. Images are unmatched if the similarity degree is 0. Another approach is based on clustering by Deng and Manjunath [6]. In this approach, a compact color feature descriptor is used to describe the image regions with sufficient number of colors and their percentage in the region. Each color in the query image or the region is used to find the regions in another image containing that color. Comparison is done using a similarity measure similar to the quadratic color histogram distance measure. For all the query colors, matching is done and then combined to obtain the final retrieval result.

8.4 Image Retrieval Model

A model for the image retrieval system is shown in Figure 8.3. For image matching, image features such as shape, texture, or color are extracted. For

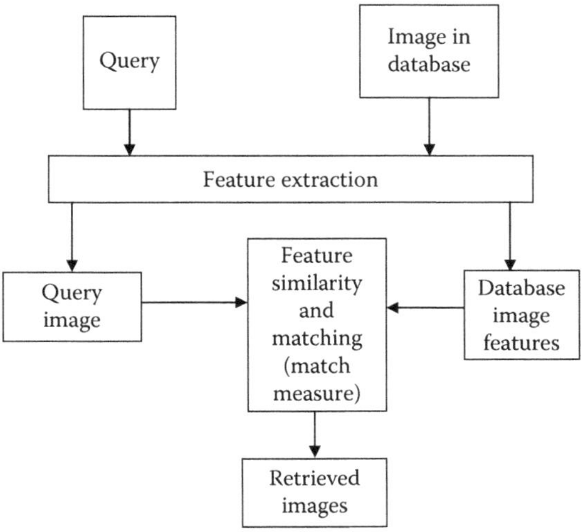

FIGURE 8.3
An image retrieval system.

each of the images in the database image, image features are computed. Another image called the query image is chosen as the input image whose features are also computed. Then the features of the query image are allowed to match with the features of each of the images in the database using some similarity or distance functions. One important criterion for testing the efficacy of search and retrieval is that the features of the query image should be present in the images in the database.

A similarity function is a mapping between a pair of feature vectors to find the visual similarity between two images. A global color histogram of an image A is represented by an N dimensional vector, $H(A)=[H(A,n), n=1, 2, 3,\ldots, N]$, where N is a number of quantization colors and $H(A,n)$ is the number of pixels having color n. The histograms (H_{AR}, H_{AR}, H_{AB}) are constructed for all the color channels, R, G, and B. The similarity between the two images is computed based on this representation. There are many similarity measures but the common types of similarity measurement are the L1 norm, L2 norm, the color histogram intersection, and weighted distance metric.

8.5 Fuzzy-Based Image Retrieval Methods

As the image pixels are imprecise and the distributions of colors are also not clear due to the quantization effects and human perception, there is an error in representing the colors. So a pixel may present in two or more bins. Fuzzy set theory may be useful in dealing with this type of uncertainty. Most of the methods for color retrieval use image histogram which is the occurrence of the gray levels in an image. But as the color, texture, or shape features in an image is not precise and clear, fuzzy histograms [4,7] and fuzzy features are modeled corresponding to the regions of an image. Fuzzy features characterize imprecise color distribution, vagueness in shape or texture in the image. The uncertainty is represented in the form of membership function, which is the degree of belongingness of the pixel in the image. Then retrieval and comparison is performed using fuzzy distance measures. We shall discuss several fuzzy methods for color retrieval.

8.5.1 Fuzzy Similarity–Based Retrieval Model

Tolias et al. [11] proposed FSMIQ, an image retrieval model that is based on fuzzy similarity metrics. The method works well in HSV color space. This technique uses color and shape information embedded in an image. In this method for efficiently encoding the color distribution information, some spatial information is used. Assuming that the object of interest is located at the center of the image (human eye mainly draws attention to the center of the image), the membership function for objects that gets its largest value in the center of the image is

$$\mu_{\text{object}} = \max\left\{\sin^2\left(\frac{2\cdot\pi\cdot i}{2\cdot M}\right)\cdot\sin^2\left(\frac{2\cdot\pi\cdot j}{2\cdot N}\right)\right\} \tag{8.1}$$

$0 \le i < M$. $0 \le j < N$, where M and N are image dimensions.

In order to represent color distribution for each color channel, c, the object and background regions are represented using the following features. The features indicate the relative distribution of pixels in histogram bin.

$$H_{\text{LOW},c} = \frac{\sum_{i=0}^{255}\mu_{\text{LOW}}(i)\cdot h(i)}{\sum_{m=0}^{M-1}\sum_{n=0}^{N-1}\mu_{\text{object}}(m,n)} \tag{8.2}$$

$$H_{\text{MED},c} = \frac{\sum_{i=0}^{255}\mu_{\text{MED}}(i)\cdot h(i)}{\sum_{m=0}^{M-1}\sum_{n=0}^{N-1}\mu_{\text{object}}(m,n)} \tag{8.3}$$

$$H_{\mathrm{HI,c}} = \frac{\sum_{i=0}^{255} \mu_{\mathrm{HI}}(i) \cdot h(i)}{\sum_{m=0}^{M-1} \sum_{n=0}^{N-1} \mu_{\mathrm{object}}(m,n)} \tag{8.4}$$

where

$$h(i) = \sum_{(m,n)=i} \mu_{\mathrm{object}}(m,n)$$

The membership functions used are LOW, MEDIUM, and HIGH as shown in Figure 8.4.

With this approach, for any color image, three features are derived for each channel for object and for background, and thus a total of 18 values are calculated using the above equations. The images are then compared using these features using HSV color model. For effectively computing the shape description of the image, discrete wavelet transform that uses Haar wavelet, is used. Tversky's index is used to carry out the similarity assessment

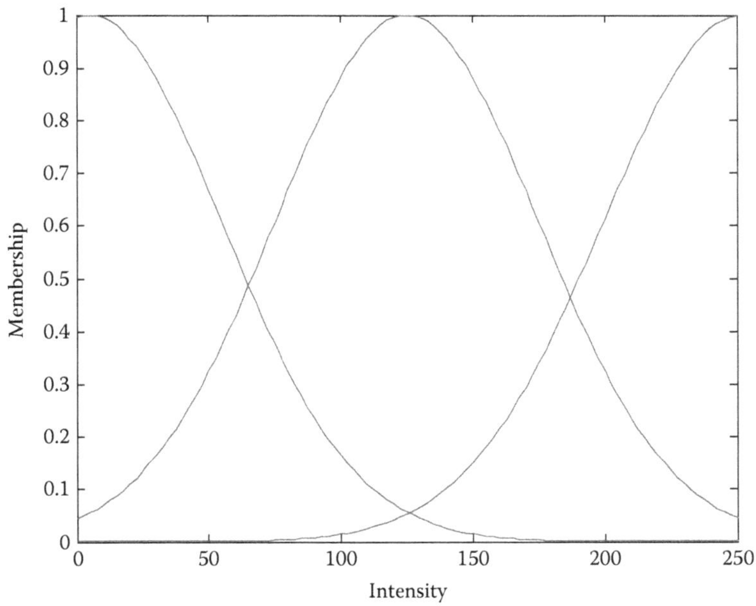

FIGURE 8.4
Membership for LOW, MEDIUM, and HIGH. (Adapted from Tolias, Y.K. et al., Fuzzy similarity for matching image queries, in *Proceedings of the International Conference on Information, Intelligent and Systems*, Bethesda, MD, pp. 259–264, 1999.)

TABLE 8.1

Retrieval Results of Flag Images Using the Methods of Vertan and Boujemaa and Tolias et al.

Methods for Color Flag Images	Precision				Recall			
	$n \leq 5$	$n \leq 10$	$n \leq 15$	$n \leq 20$	$n \leq 5$	$n \leq 10$	$n \leq 15$	$n \leq 20$
Vertan and Boujemma	0.84	0.67	0.55	0.46	0.35	0.55	0.63	0.69
Tolias et al.	0.54	0.40	0.32	0.28	0.24	0.30	0.35	0.45

by aggregating the fuzzy features. Results are obtained separately for color retrieval and shape retrieval and final retrieval is obtained after combining the color and shape retrieval.

A color retrieval result is shown in Table 8.1.

8.5.2 Color Histogram–Based Retrieval

Vertan and Boujemma [12] used color histograms and distance measures for fuzzy color retrieval. Initially the histogram is normalized with the maximum value in the histogram bin. For a color image, if c is the color at location (i, j) i.e., $c=f(i, j)$, the color distribution of the color set ξ is

$$h_1(c) = \frac{h(c)}{\max_{c \in \xi} h(c)}, \quad \forall c \in \xi \tag{8.5}$$

This normalization mode assigns the maximum to the color that is more dominant in the image. Then using a distance measure, L1 metric, between the query image and the images in the database, similar images are retrieved.

8.5.3 Smoothed Histogram–Based Retrieval

Vertan and Boujemaa [12] also used another type of color retrieval using a smoothed fuzzy histogram put forward by Jawahar and Ray [7]. Fuzzy histogram takes into account a gray level g that can be either $g+1$ or $g-1$. The computation of color histogram is based on the principle that a color is uncertain due to the erroneous version of another color. Thus a pixel of any color will not only contribute to its own bin but also to its neighboring bin. This amount of uncertainty is measured using a triangular membership function. This technique is a reformulation of the classical histogram in Equation 8.5 where the usual histogram is convolved with a Gaussian kernel for smoothing as suggested by Kaustsky et al. [8]:

$$h_2(c) = h(c) * k(c) \tag{8.6}$$

Every color c is modeled as a fuzzy set C. For every possible color c' of the color space, the membership function to the fuzzy set C is $\mu_C(c')$ that represents the credibility or the similarity that the color c' may represent c. The membership function is defined as

$$\begin{aligned}\mu_C(c') &= 1, d < \text{JND} \\ &= \exp\left(-\frac{d/\text{JND}-1}{\lambda^2}\right)^2, \quad \text{otherwise}\end{aligned} \tag{8.7}$$

where d is the inter-color Euclidean distance, $d = |c - c'|$ and JND is a user defined. If the color space is CIELAB color space. λ is a parameter that controls the credibility f the color pair and is chosen adaptively.

Finally, the fuzzy histogram aggregates by a T-conorm σ [13], the credibilities of the approximation of the color c' with respect to color c of each pixel of the image into the credibility with respect to the entire image. The fuzzy histogram is represented as

$$h_3(c') = \sigma_{c \in \xi}(\mu_C(c')), \quad \forall c' \in \xi \tag{8.8}$$

The particular nature of the fuzzy histograms h_3 use L1 metric only but the histogram h_2 uses any metric to find the similarity of colors between the two images. Given a query image and a database image, similarity is calculated between the query image and each of the images in the database for all the three color channels. Then total similarity value is computed after adding the similarity for the three color channels.

Example 8.1

Retrieval results of Vertan and Tolias et al. are shown in Table 8.1. Vertan's method on the normalized histogram h_1 in Equation 8.5 using L1 metric is shown. Normalized histograms are computed for both the query image and the images in the database. Using the L1 metric, similarity between the query image and each of the images in the database is computed for each color channel. Tolias's method on color image retrieval is shown.

The major objective of retrieval is to develop efficient image retrieval, which is insensitive to orientation. Retrieval is defined as accurate if for a given query image, the most similar image in the database is retrieved by the system and is marked as the topmost retrieval. It is desirable to have an efficient retrieval scheme, since image databases usually have a large set of images.

The test is performed on the database of flag images with 100 flags of different countries obtained from the site: http://www.countryreports.org/flags/nationalflags.aspx. Ten images are picked randomly from the image database and each is presented as a query image. In this experiment, only few colors are taken into account. Retrieval statistics is computed using the colors in the images in the database with specified set of colors present in the query. For each of the query image, the first 20 retrieved images are listed for each query image using the color retrieval method. Also the images in the database that are similar to the query image are listed. Precision and recall rates [3] for measuring the accuracy of the retrieval scheme are calculated at positions 5, 10, 15, and 20. If n_1 is the number of images retrieved in the top 5, 10, 15, and 20 positions that are closed to the query and n_2 is the number of images in the database similar to the query, then

Recall rate $= n_1/n_2$

Precision rate $= n_1/n$, $n = 5, 10, 15$, or 20.

In the retrieval result in Table 8.1, the average of the 20 query precision/recall values are taken in calculating the precision and recall.

A result on flag images (Vertan and Boujemaa) is shown in Figure 8.5.

8.5.4 Fuzzy Similarity/Tversky's Measure–Based Retrieval Method

We now discuss another fuzzy color retrieval method [1,4] that uses several fuzzy similarity measures. The idea is to find out the image in the database that is the most similar to the query image. *R, G, B* color model is used in this method.

Consider two RGB color images *A* and *B* of size $M \times M$. Three color histograms R_{An}, G_{An}, and B_{An} of each image *A* in the database are computed. In a similar manner, for the query image *B*, three color histograms R_{Bn}, G_{Bn}, and B_{Bn} are also computed. Though there are many ways one can define the membership functions, here gamma membership function (as discussed in Chapter 5) is used in the color retrieval method and is as follows:

$$f(x) = \exp(-(x - \nu)) \tag{8.9}$$

The membership values for each of the color channels *R*, *G*, and *B* are computed using gamma membership function. The location parameter ν is the histogram of the query image. *A* is an image in the database and image *B* is

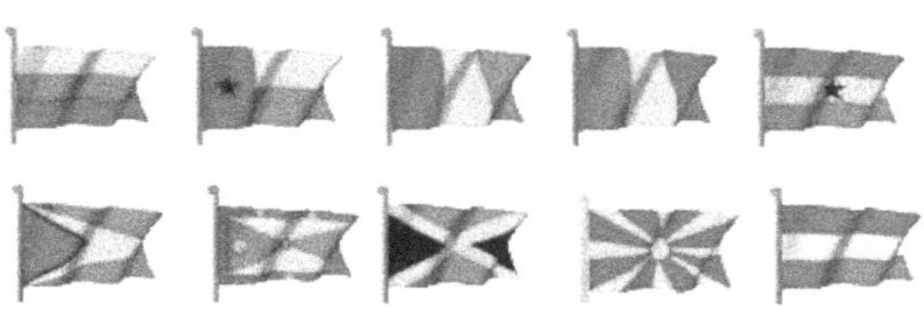

FIGURE 8.5
(See color insert following page 144.) Sample results of flag images. (Courtesy of www.CountryReports.org)

the query image, which may or may not be part of the database image. Thus ν is replaced by h_B, the histogram of the query image, that is, the membership values of the images in the database are computed with respect to the query image or the reference image. So the membership function of each image in the database image for each color is

$$\mu_{A(R,G,B)}(n) = \exp\left(-\text{const}_{(R,G,B)} * \text{abs}\left(h_{A(R,G,B)}(n) - h_{B(R,G,B)}(n)\right)\right) \tag{8.10}$$

and the constant is calculated as

$$\text{const}_{(R,G,B)} = 1 \Big/ \left[\max_k \left\{\max\left(h_{A(R,G,B)}(k)\right)\right\} - \min_k \left\{\min\left(h_{A(R,G,B)}(k)\right)\right\}\right], \quad k = 1, 2, 3, \ldots, K \tag{8.11}$$

where
K is the total number of images in the database
h_A is the histogram of each image in the database

The membership values of the query image will be 1.

$$\mu_{B(R,G,B)}(n) = \exp\left(-\text{const}_{R,G,B}\left(h_{B(R,G,B)}(n) - h_{B(R,G,B)}(n)\right)\right) \tag{8.12}$$

$n = 0, 1, 2, \ldots, L-1$, is the gray levels of the image. The constant is used to limit the membership value feasible in the interval [0,1].

In words, the constant is computed as follows: initially the maximum and minimum counts of all the histograms of the images in the database are noted. If there are K images in the database, there will be K maximum values and K minimum values. The maximum value amongst these K numbers of maximum values is selected and the minimum value amongst the K minimum values is selected. Then the constant is the inverse of the difference between the maximum and the minimum values. By doing so, the most probable color will have a membership degree 1 within a fuzzy image.

The calculation of constant is described in flowchart in Figure 8.6.

For each image in the database and the query image, the membership function is calculated. Then using fuzzy similarity measures, the retrieval result is computed. We shall describe a set of interesting fuzzy similarity measures that are used to find the similarity between the query image and each of the images in the database.

8.5.4.1 Fuzzy Similarity Measures

8.5.4.1.1 Similarity Measure Based on Min–Max Ratio

The similarity between two images A and B is given by

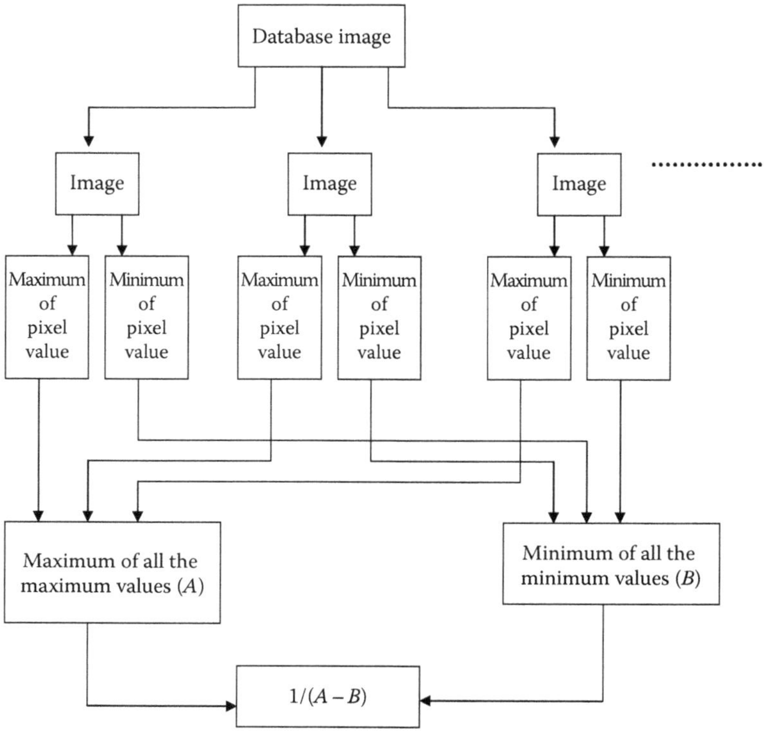

FIGURE 8.6
Block diagram of constant calculation.

$$S_1(A,B) = \frac{\sum_{n=1}^{L} \min(\mu_A(n), \mu_B(n))}{\sum_{n=1}^{L} \max(\mu_A(n), \mu_B(n))} \tag{8.13}$$

where $\mu_A(n)$ and $\mu_B(n)$ are the membership values of the nth gray level of the histograms A_n and B_n of the two images A and B, respectively. For an exactly identical pair of images, the similarity value will be equal to 1.

8.5.4.1.2 Similarity Measure Based on Contrast De-Enhancement

Using Wang's formula for similarity, the contrast de-enhancement-based similarity between two images A and B is

$$S_2(A,B) = (1/n) * \sum_{n=1}^{L} (1 - |\mu_A(n) - \mu_B(n)|) \tag{8.14}$$

It is also almost the same as the previous one, where for exactly an identical images, the value is 1.

$$\text{Since } (1/n)^* \sum_{n=1}^{L} \left(1 - |\mu_A(n) - \mu_B(n)|\right) = 0 \quad \text{as } \mu_A = \mu_B$$

8.5.4.1.3 Similarity Measure Based on Normalized Absolute Difference

This is another similarity measure that is based on the normalized absolute difference between the membership values of the gray levels computed from the histograms of the two images. This measure is also observed to yield appropriately similar images from the database images and is as below

$$S_3(A,B) = 1 - \frac{\sum_{n=1}^{L} \text{abs}(\mu_A(n) - \mu_B(n))}{\sum_{n=1}^{L} (\mu_A(n) + \mu_B(n))} \tag{8.15}$$

8.5.4.1.4 Measure Based on Fuzzy Divergence

The total fuzzy divergence between the images A and B computed from the histograms A_n and B_n, respectively, is given as

$$\begin{aligned} S_4(A, B) &= \sum_{n=0}^{L-1} (2 - (1 - \mu_A(n) + \mu_B(n))e^{\mu_A(n) - \mu_B(n)} \\ &\quad - \left(1 - \mu_B(n) + \mu_A(n)e^{(\mu_B(n) - \mu_A(n))}\right) \end{aligned} \tag{8.16}$$

The less is the divergence, the most similar the image is, i.e., the similarity is inversely proportional to the divergence. For two identical images the divergence value is 0.

8.5.4.1.5 Measure Based on Fuzzy Set Theoretic Approach

This similarity measure based on Generalized Tversky's index (GTI) is described in Chapter 5. The GTI, as discussed in Chapter 3, is written as

$$\text{TIM}(A,B;\alpha,\beta) = \frac{\sum_{n=1}^{L} \min(\mu_A(n),\mu_B(n))}{\sum_{n=1}^{L} \left(\begin{aligned} &\min(\mu_A(n),\mu_B(n)) + \alpha \cdot \min(\mu_A(n), 1 - \mu_B(n)) \\ &+ \beta \cdot \min(1 - \mu_A(n), \mu_B(n)) \end{aligned} \right)} \tag{8.17}$$

The more the GTI between the two images is, the more is the common feature and less is the distinctive features. This implies that the two images are more similar images. The values α=0.5 and β=0.5 are used.

Fuzzy similarity measure is computed between each of the images in the database and the query image. The total similarity for all the color channels between an image in the database and the query image is $\sum_{R,G,B} S$, where S is the fuzzy measures S_1, S_2, S_3, S_4, and GTI model (Equations 8.13 through 8.17). The similarity is calculated for all the images in the database. The image whose similarity value is maximum or minimum, depending on the type of similarity measure, is the most similar to the query image. The images in the database that are most similar to the query image is retrieved and the retrieval percentage is calculated for the top 20 retrieved images that are almost similar to the query image.

Algorithm: 8.1 Fuzzy color retrieval using fuzzy similarity measure

Step 1. Compute the fuzzy feature (here fuzzy color histogram) for the query image.
Step 2. Compute the similar fuzzy feature for each for the images in the database.
Step 3. Apply fuzzy similarity measure between the query image and each of the images in the database for a channel.
Step 4. Repeat steps 1–3 for the other channels.
Step 5. Combine the similarity measures for all the channels.
Step 6. Select the top 20 retrieved images that are similar to the query.
Step 7. Select 20 similar images from the database.
Step 8. Compute the retrieval efficiency.

TABLE 8.2

Retrieval Results of Color Flag Images

	Precision				Recall			
Methods	$n \leq 5$	$n \leq 10$	$n \leq 15$	$n \leq 20$	$n \leq 5$	$n \leq 10$	$n \leq 15$	$n \leq 20$
Divergence								
Normal	0.77	0.64	0.51	0.43	0.33	0.55	0.60	0.69
Rotated	0.82	0.65	0.50	0.44	0.36	0.55	0.64	0.73
Similarity measure (S_1)								
Normal	0.85	0.70	0.57	0.49	0.38	0.60	0.69	0.77
Rotated	0.83	0.74	0.57	0.46	0.35	0.62	0.70	0.78
GTI								
Normal	0.87	0.69	0.56	0.48	0.40	0.60	0.69	0.75
Rotated	0.90	0.70	0.57	0.49	0.39	0.60	0.70	0.75

Example 8.2

We will show an example of a retrieval result on flag images for different countries for Chaira and Ray [4]. The database of flag images contains 100 flags of different countries. Twenty images are picked randomly from the image database and each is presented as a query image. As the method is histogram based, the result is rotation invariant. To show in practice the rotation invariancy, retrieval percentage is shown on rotated images. Retrieval calculation is similar to the one described in Section 8.5.3.

Table 8.2 shows the precision and recall accuracy on color flag using all the measures. Figure 8.7 shows the sample results of ranking the images

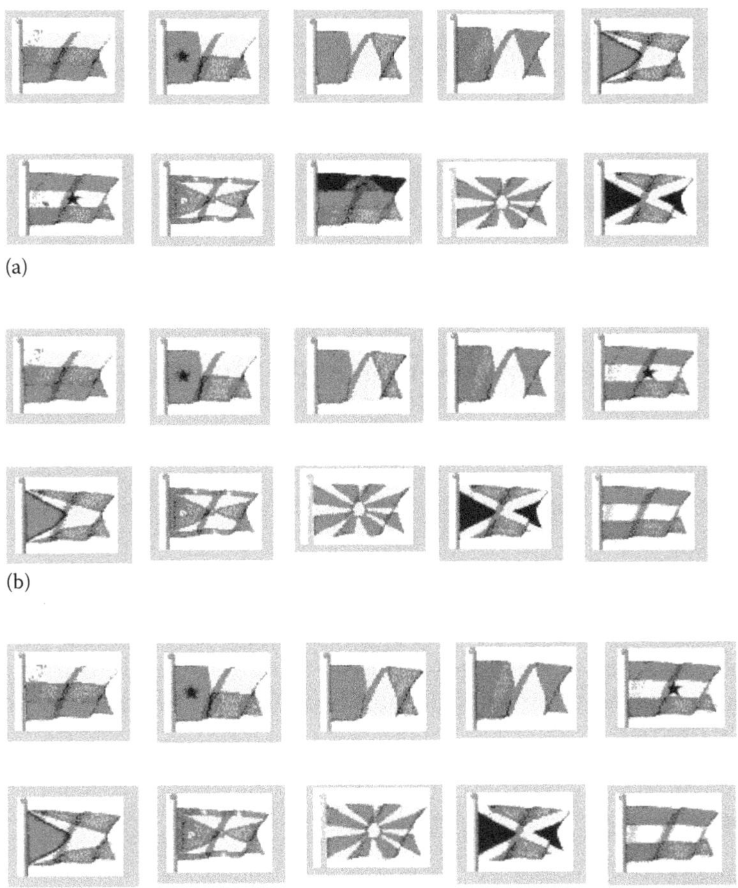

(a) (b) (c)

FIGURE 8.7
(See color insert following page 144.) Sample results of normal flag images using (a) fuzzy divergence, (b) similarity measure, and (c) GTI model.

in the database according to the query image where the query image is at the extreme top left of the figures (marked as query). Top ten images are displayed according to the similarity of colors present in the query image in descending order, that is, the most similar image is at the top.

Figure 8.7a through c displays the results using fuzzy divergence, fuzzy similarity measure, and GTI model respectively of normal flag images.

8.6 Summary

Content-based image retrieval is discussed here. In content-based image retrieval, given a query image and a database image, the image in the database that is most similar to the query image is retrieved. The features that are used to find the similarity of images are color, texture, or shape. Retrieval can be carried out by considering the image as crisp or fuzzy. Fuzzy color retrieval is also described.

There are different approaches proposed by several authors for color image retrieval. Based on the type of requirements by the user, the approaches are categorized in three classes:

1. Global based—It takes into account the histogram of the image (color feature) and the histograms are used to retrieve the images.
2. Partition based—The image is divided into a set of sub-windows. The features of each sub-window of the two images (query image and an image in the database) are used to retrieve the query image in the database.
3. Region based—The image is divided into regions and the features of the region are compared.

As the image pixels are imprecise and the distribution of color is not clear, fuzzy approach is used. Few fuzzy methods are discussed that use fuzzy histogram using different membership functions and retrieval is performed using fuzzy similarity measures. Some methods use fuzzy features to compare the two images. For image comparison, several fuzzy similarity measures are discussed.

References

1. Acharya, T., Chaira, T., and Ray, A.K., Image color matching scheme, Intel US Patent 7133555, 2006.
2. Ballard, D.H. and Brown, C.M., *Computer Vision*, Prentice Hall, Inc., Englewood Cliffs, NJ, 1982.

3. Bimbo, A.D., *Visual Information Retrieval*, Morgan Kaufmann Publishers, Inc., San Francisco, CA, 1999.
4. Chaira, T. and Ray, A.K., Fuzzy measures for color image retrieval, *Fuzzy Sets and Systems*, 150(3), 545–560, 2005.
5. Chen, Y. and Yang, J.Z., A region based fuzzy feature matching approach to content based retrieval, *IEEE Transactions on Pattern Analysis and Machine Intelligence*, 24(9), 1252–1267, 2004.
6. Deng, Y. and Manjunath, B.S., An efficient color representation for image retrieval, *IEEE Transactions on Image Processing*, 10(1), 140–147, 2001.
7. Jawahar, C.V. and Ray, A.K., Fuzzy statistics of digital image, *IEEE Signal Processing Letters*, 3(8), 225–227, 1996.
8. Kaustsky, J., Nichols, N.K., and Bupp, D.L.B., Smoothed histogram modification in image processing, *CVGIP*, 26(3), 271–291, 1984.
9. Li, Y., Wan, X., and Jay Kuo, C.-C., Introduction to content based image retrieval-overview to key techniques, in *Image Database*, Eds.: Vittorio Castelli, Lawrence D. Bergman, John Wiley and Sons, 2002.
10. Sonka, M. et al., *Image Processing, Analysis and Machine Vision*, Brooks/Cole, Thomson Learning, Pacific Grove, CA, 2001.
11. Tolias, Y.K. et al., FSMIQ: Fuzzy similarity for matching image queries, in *Proceedings of the International Conference on Information, Intelligent and Systems*, Bethesda, MD, pp. 259–264, 1999.
12. Vertan, C. and Boujemma, N., Use of fuzzy histogram and distances for color image retrieval, in *Challenges of Image Retrieval*, Brighton, U.K., pp. 1–6, 2000.
13. Zimmerman, H.J., *Fuzzy Sets, Decision Making and Expert Systems*, Kluwer Academic Publisher, Boston, MA, 1997.
14. http://www2.ncsu.edu./scivis/lessions/colormodels/# HSV

FIGURE 6.7a
Scene image "zebra".

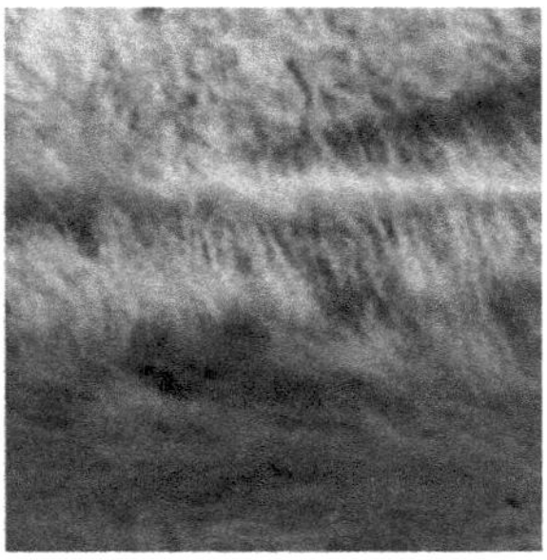

FIGURE 6.8a
Scene image "grassland".

FIGURE 6.9a
Scene image "doc cage city".

FIGURE 7.10a
"Lena" image.

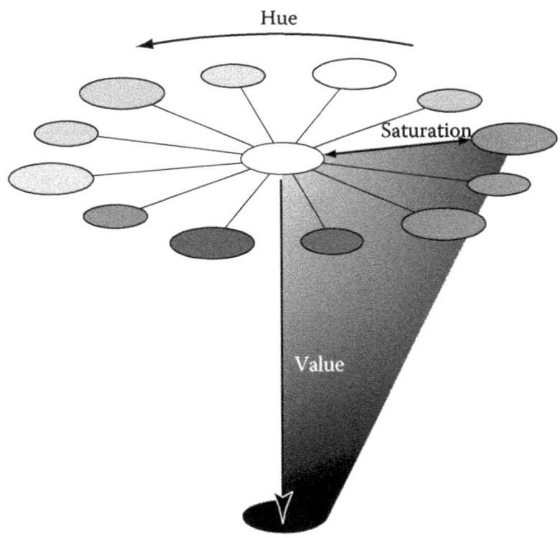

FIGURE 8.1b
HSV model. (http://www.ncsu.edu.scivis/lessons/colormodels/color_modelse.html. With permission.)

FIGURE 8.5
Sample results of flag images. (Courtesy of www.CountryReports.org)

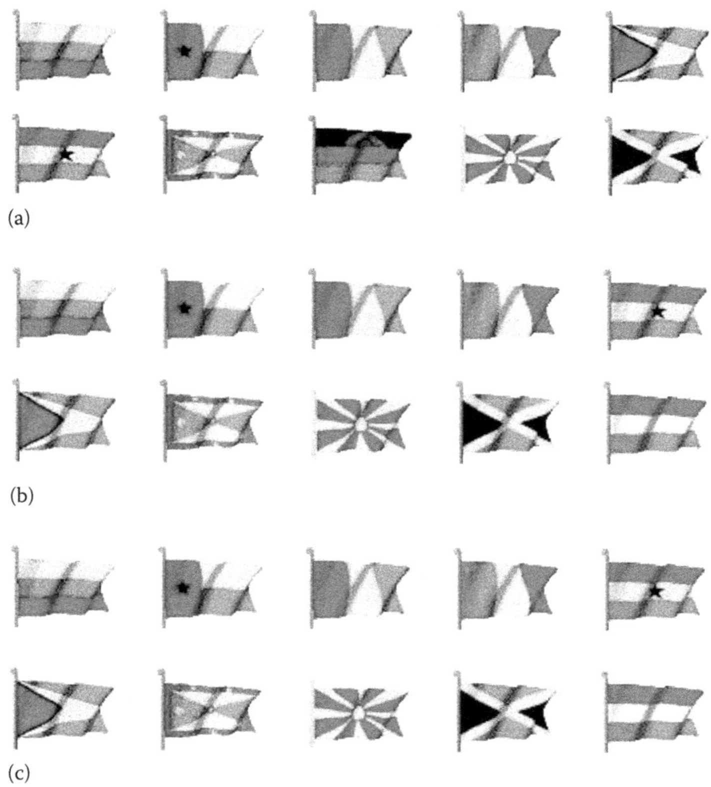

FIGURE 8.7
Sample results of normal flag images using (a) fuzzy divergence, (b) similarity measure, and (c) GTI model.

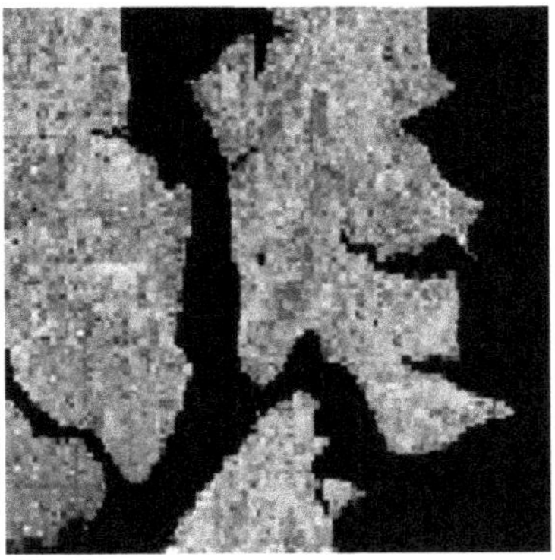

FIGURE 10.1a
Remote sensed image.

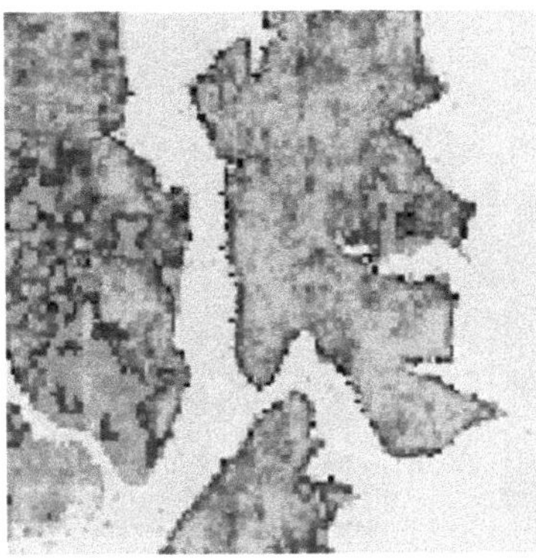

FIGURE 10.1d
Image of (a) in HSV model.

FIGURE 10.2a
Remote sensed image.

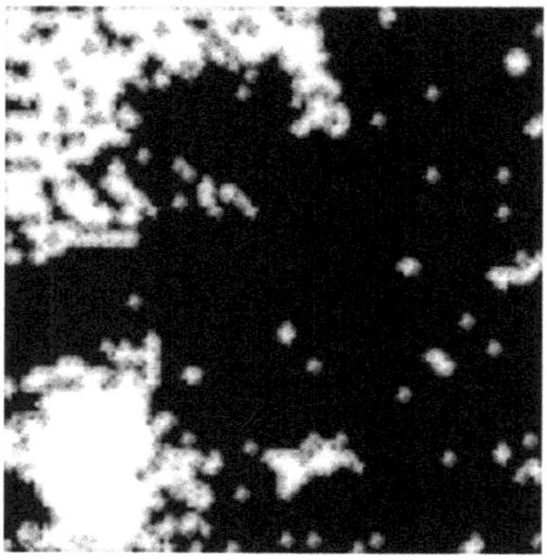

FIGURE 10.2b
Region-extracted image of (a) in RGB model.

FIGURE 10.2c
Query image.

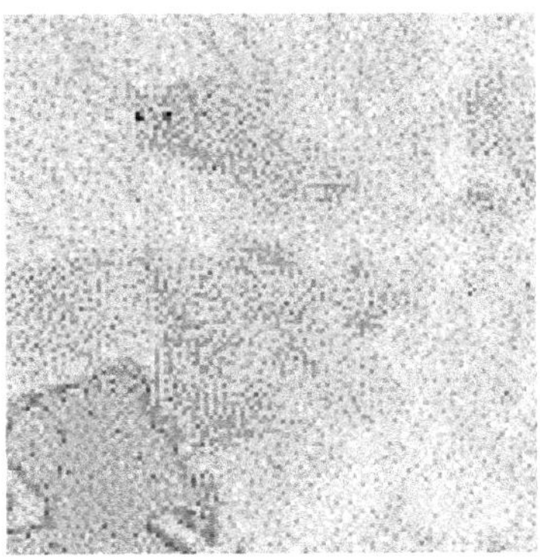

FIGURE 10.2d
Image of (a) in HSV model.

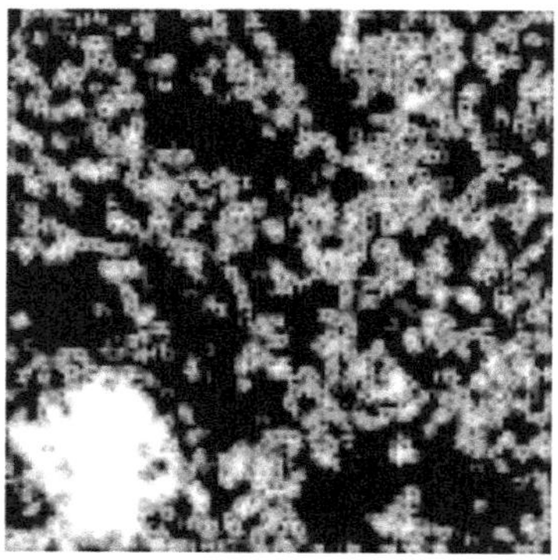

FIGURE 10.2e
Region extracted using HSV model.

9

Fuzzy Methods in Pattern Classification

9.1 Introduction

In a pattern, the information of interest to an observer can be characterized in terms of some basic properties of the perceived pattern. The analysis of these properties involves the measurement of a set of variables. The first stage of acquisition of pattern data together with all the subsequent stages of complex cognitive processing, leading to the final interpretation of the pattern, is called pattern recognition.

The tasks as outlined above can be selectively specified as the problem of (a) pattern sensing, (b) pattern characterization, and (c) abstraction and generalization.

The problem of sensing involves the design of a device for detection and acquisition of the pattern. This is followed by the task of noise cleaning and representation of pictorial or non-pictorial (such as sound, earthquake, medical) form of pattern on a computer. The problem of characterization is one of associating a set of attributes or features with each pattern under consideration. Once the features are extracted, the space of these measured feature values is partitioned into a set of meaningful regions in the classification phase. This is done using broadly two approaches: (a) decision theoretic approach and (b) linguistic approach. In both these approaches again the classification algorithms can be deterministic, statistical, or fuzzy.

In the decision theoretic approach, each pattern is characterized by a set of measured features and thus each pattern can be viewed as a point in the multidimensional feature space.

In the classification phase, based on suitable criteria, the data space corresponding to the pattern is partitioned in several regions. This is precisely the problem of pattern classification, where search is made for structure in the pattern space. The classifier is designed with the help of sample patterns. As mentioned, depending on the nature of the problem, the classification techniques can be deterministic, statistical, or fuzzy. Thus the fuzzy techniques are employed in situations where the feature vectors under consideration are essentially fuzzy in nature. Once the partitioning in several decision regions is over, the suitability of the decision function is assumed for classification of patterns not included in the sample set—this is termed as generalization.

A large number of research findings have been reported on various facets of Pattern classification. One can find excellent ideas from some of the historic books, such as Duda and Hart [8], Tou and Gonzalez [20], Fu [11], Majumdar and Ray [14], and also from the more recent ones, such as, Duda et al. [7], Shawe-Taylor and Cristianini [17], Friedman and Kandel [10], Theodoridis and Koutroumbas [19], Fukunaga [12], Kohonen [13], Bishop [4], and so on. In Section 9.2, we discuss the fundamentals of decision theoretic pattern classification.

9.2 Decision Theoretic Pattern Classification Techniques

The foundation of decision theoretic pattern recognition involves feature extraction, classification, and learning as described in Figure 9.1.

In deterministic techniques, the patterns are represented as n-dimensional feature vectors and decision making is formulated in terms of discriminant functions. The classifier is expressed in terms of a set of m number of discriminant functions $g_i(x)$, $i=1, \ldots, m$, for m-class problem and a feature vector x is assigned to class C_i, if $g_i(x) > g_j(x)$ for all $j \neq i$.

Out of many linear and nonlinear discriminant functions reported, the simplest one is a linear discriminant function, which is of the form

$$g_i(x) = \sum_{k=1}^{n} w_{ik} x_K + w_O \quad \text{for } i = 1, 2, \ldots, m \tag{9.1}$$

where

w_{ik}'s are weighting parameters

w_O is the threshold weight

The equation $g_i(x)=0$ defines the decision surface that separates the samples assigned to class C_i from other classes. The decision surface becomes a hyperplane if $g_i(x)$ is linear. Thus a set of m linear discriminant functions can be used to divide the feature space in m decision regions and the resulting classifier is known as a linear machine. The set of decision regions for a

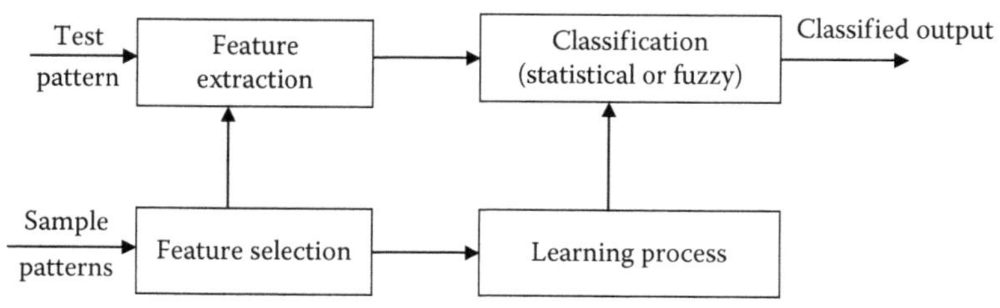

FIGURE 9.1
Decision theoretic pattern classification.

linear machine are convex. In cases where the patterns are not linearly separable, generalized nonlinear discriminant functions have been suggested.

One such nonlinear discriminant function is quadratic discriminant function, of the form

$$g_i(x) = \sum_{j=1}^{d}\sum_{k=1}^{d} w_{ijk}x_j x_k + \sum_{k=1}^{n} w_{ik}x_k + w_O \quad \text{for } i = 1, 2, \ldots, m. \tag{9.2}$$

As may be observed the quadratic discriminant function yields a hyper quadratic separating surface between each pattern class.

In practice, the weighting parameters for the discriminant functions are not known a priori. A set of labeled training patterns are used for estimation of the weighting parameters and the efficiency of the classifier is enhanced as a finite large number of training samples from each pattern class are made available to the classifier. A set of pattern samples are said to be linearly separable, if there exists a weight vector that classifies all the samples correctly.

For computing the set of linear discriminant functions, various learning strategies, namely, gradient descent procedure, perceptron criterion, least mean square (LMS) criterion, relaxation procedures, and so on have been suggested. The statistical pattern classification techniques deal with further generalization of the discriminant function approach. Here the unknown parameters of the discriminant functions are estimated utilizing the statistical information of the pattern classes. One of the most widely used approaches to statistical pattern classification using the Bayesian rule can be stated as follows.

Let us assume that the a priori probability for the occurrence of the class C_j be $P(C_j)$. Also let the state conditional probability density function $p(x|C_j)$ be known, then the a posteriori probability of the pattern sample x belonging to the class C_j is given by

$$P\left(C_j|x\right) = \frac{p\left(x|C_j\right)\cdot P\left(C_j\right)}{\sum_{i=1} p(x|C_i)P(C_i)} \tag{9.3}$$

The question that may arise is, "what should be the appropriate model for the density function $p(x|C_i)$ and what will be the parameters of that model?" When the feature vector x for a given class C_j is a mildly corrupted version of the prototype vector μ, then we can choose the multivariate normal density as an appropriate model for the density function. The multivariate normal density can be specified completely by $n+n(n+1)/2$ parameters, which includes n number of elements of the mean vector μ and a set of $n(n+1)/2$ number of elements of the covariance matrix Σ. The mean vector μ and the covariance matrix Σ can be estimated using supervised learning approach. In supervised learning approach, a large number of labeled training pattern samples are used to estimate the mean vector μ and the covariance matrix Σ. The a posteriori probability density function $p(x|C_j)$ can be computed recursively and the computational feasibility of these recursive estimation algorithms depend on the sufficient statistics of the density function. It has been shown that when the probability density function is Gaussian in nature, the recursive estimation of the a posteriori probability is equivalent to the Kalman–Bucy filtering problem.

In many cases, the assumption that the pattern data comes from Gaussian distribution may not be a valid one. Alternatively in these cases a number of different parametric forms of the underlying probability distribution can be considered and the best fit yields a better estimate of the distribution. In supervised classifiers, the estimation of the condition probability density functions are achieved using computationally simple algorithms from a large number of labeled patterns. A wrong choice of parametric form may, however, sometimes result in higher probabilities of misclassification.

In many practical applications, the number of pattern classes and dimensionality of the feature space can be large enough to render a single-level classification procedure difficult. Hierarchical decision trees have been suggested for efficient pattern classification in these cases.

The ideas of minimum distance classifier can be extended to the problems where prototypes may not be available and nearest neighbors classification rules can be applied in these situations. In nearest neighbor classification, an unknown sample is classified to the category to which its nearest neighbor from the sample set belongs. The idea of 1-N classification rule can further be extended to the K-NN rule where the decision about the class membership of the unknown sample is taken by considering the majority votes amongst the K nearest neighbors. It can be shown that the error performances of the NN classification rule are excellent. However, the probability of error is bounded by twice the Bayesian error rate.

In many classification problems, a set of correctly labeled patterns may not be available and the available pattern samples may be unlabeled. In such cases we cannot use supervised classification schemes. The unsupervised classification schemes, which are widely used in such cases, will be provided in detail in the next section.

9.2.1 Preliminaries of Unsupervised Classification

The problem of clustering is one of partitioning a set $\{x_1, \ldots, x_n)$ of elements or samples into a finite number of subsets $X_1, \ldots, X_m$ such that each subset X_i has a common attribute in all its elements but is dissimilar from other subsets $X_j (j \neq i)$. Here each of the subsets X_i, $1 \leq i \leq m$ is known as a cluster. The similarity measure is the basic building block of any clustering criteria.

It is necessary to define a measure of similarity or distance existing between the elements of the set to be clustered.

Some of the metric functions which can be used to compute the distance or the "similarity" are listed below.

Name **Form of Metric Function**

Euclidean

$$d_2(x_i, x_j) = \left[\sum_{k=1}^{p} (X_{ki} - X_K)^2 \right]^{1/2}$$

G norm

$$d_1(X_i, X_j) = \left[\sum_{k=1}^{p} |X_{ki} - X_{kj}| \right]$$

Supervised norm

$$d_\alpha(X_i, X_j) = \left[k = 1, p \overset{\text{sup}}{|X_{ki} - X_{kj}|} \right] \tag{9.4}$$

L norm

$$d_p(X_i, X_j) = \left[\sum_{k=1}^{p} |X_{ki} - X_{kj}|^p \right]^{1/p}$$

Mahalanobish

$$D^2(X_i, X_j) = (X_i - X_j)^T W^{-1} (X_i - X_j)$$

The supervised norm is computationally simple but involves a ranking procedure. Mahalanobish metric is often referred to as a generalized Euclidean distance. The matrix W^{-1} denotes usually the inverse of scatter matrix.

$$W = \sum_{i=1}^{n} (X_i - \bar{X})^2 = \sum_{i=1}^{n} X_i^2 - \frac{1}{n}\left(\sum_{i=1}^{n} X_i \right)^2 \tag{9.5}$$

where X_i is the ith sample.

Mahalanobish distance is invariant to any nonsingular linear transformation.

Interestingly there does not exist a unique clustering algorithm which will be universally applicable for all situations. The choice of a particular clustering criterion has always been a subjective phenomenon and depends on the structure of the data set. Clustering techniques are suggested by the different authors as Bezdek [1,3] and Dubes [6]. The various types of clustering criteria can broadly be grouped as (a) objective functional, (b) hierarchical, and (c) graph theoretical.

The solution to clustering problem using objective functional method determines a partitioning satisfying an optimality criterion given in terms of a functional relation known as objective functional. The set of objective functionals are chosen which provide a suitable measure of the clusters and the local extreme of the objectives functional identifies with optimal clustering. A number of objective functions have been used by the researchers for ensuring that partitioning is appropriately performed. For example, a good clustering technique must ensure that there is little variance within the data set belonging to a cluster and at the same time between the class variance amongst the classes should be maximized. Thus we can say that within the group sum of square (WGSS) errors between the samples X_is and the cluster center $\bar{X}$ where

$$\text{WGSS} = \left(X_i - \bar{X}\right)^2 = \sum_{i=1}^{n} X_i^2 - 1/n\left(\sum_{i=1}^{n} X_i\right)^2 \tag{9.6}$$

provides a good function for optimal clustering.

In many situations, where there is a large number of pattern samples, the computational complexity of the clustering procedures can be reduced to a large extent if the subsets X_i are determined in several sequential steps, thus generating a hierarchy of nested clusters. This is the problem of hierarchical clustering. Hierarchical clustering methods can again be grouped into either divisive (splitting) or agglomerative (merging) methods.

In divisive techniques, the entire set of samples is partitioned in n subsets whereby the interclass distance is maximized and this dichotomy continues till the algorithm terminates according to some optimality criteria. In agglomerative clustering, on the other hand, the procedure starts with a set of singleton elements, and the set of most similar elements at the $(p-1)$th step are merged into a larger subset in the pth step. The process of merging the similar elements continue till all the clusters are identified, in accordance with some chosen criteria.

The concept of similarity has been used for nonparametric clustering based on the sharing of near neighbors. Graph theoretical approach using the concept of n-connectedness of graphs designating the number of paths between them, has been suggested. Various hierarchical approaches to graph theoretic method of clustering include single link clustering and edge inconsistency method.

The question regarding the tendency of data to cluster poses a major problem in clustering analysis. The basic question to all the clustering methods concerns the selection of the number of classes or clusters—this is the cluster validity problem.

An important problem confronted in designing a classifier, both parametric and nonparametric is the effect of high dimensionality of the data set. If the number of features is high, then the estimation of the parameters becomes computationally very expensive. To take care of the so-called curse of higher dimensionality, various approaches have been suggested. These include the classical principal component analysis and factor analysis for reducing the dimensionality by forming linear combinations of features. A number of easily computable criterion functions, for example, entropy criterion, divergence criterion, and tr $SW^{-1}S$ criterion have been proposed for dimensionality reduction by combining the features in a linear fashion or by selecting a good subset of the available features.

9.3 Why a Fuzzy Classifier?

The importance of designing supervised and nonsupervised fuzzy classifiers cannot be overemphasized. Here we will first discuss about the limitations of statistical classifiers, followed by various fuzzy set theoretic techniques in pattern classification.

9.3.1 Limitations of Statistical Classifiers

In the light of the above discussion on various statistical classification techniques, the major limitations of statistical techniques can be outlined as below.

1. The statistical techniques can very well be applied if the functional form of the conditional density functions is known a priori and a large number of training samples are available for estimation of the parameter vector.
2. However, common parametric forms rarely fit the densities actually encountered in practice. Even if the functional form of density function is known, sufficiently large set of training samples is required for the parameter estimation. This may not be available in many practical applications.
3. If the functional form is known to be Gaussian, the problem of estimation of the distribution function is a tractable problem. Thus when sufficient statistical information regarding the pattern class is not available, use of statistical techniques for designing a suitable classifier can be a complex problem.
4. Quite often the nature of imprecision in pattern sample set is such that they are more fuzzy than stochastic in nature. Thus applying statistical techniques in such cases may not yield appropriate results.

5. Various cognitive and decision-making processes can be viewed as fuzzy in nature, since in these processes the final abstraction regarding the cognitive classification involves gradations of membership values of the objects in each class.
6. The basic human reasoning process is not traditional, two-valued, or even multi-valued logic but it is logic with fuzzy paths, fuzzy connectives, and the fuzzy rules of inference.

Thus in many pattern recognition problems, the knowledge regarding the pattern classes is imprecise in the sense that there does not exist well-defined criteria of pattern membership assignment. Moreover, a large number of pattern samples may not be available for estimation of statistics. In all these cases fuzzy approaches to pattern classification have found applications. The essence of all fuzzy classification approaches is that the classification algorithm yields a membership value of each pattern to a pattern class, which is fuzzy in nature. In the next section, fuzzy approaches to pattern classification will be briefly discussed.

9.4 Fuzzy Set Theoretic Approach to Pattern Classification

Since fuzzy set theory was first introduced by Zadeh [21], a large number of applications on fuzzy set theoretic pattern classification approaches have been reported. A large body of literature is available which deal with various fuzzy operations and processes. Fuzzy set theory has been suggested to handle the imprecision associated with real-world events by assuming that the truth value of a proposition lies in between 0 (absolute false) and 1 (absolute true). Thus a fuzzy subset A of $X=\{x\}$ which is a collection of objects, can be defined as a set of ordered pairs

$$A=[x,\mu(x)], \quad x\subset X \tag{9.7}$$

Where, $\mu_A(x)$ is the membership grade associated with x in A. In case μ: $X\rightarrow M$, which is a mapping from X to a membership space M, if M assumes values 0 and 1, the set becomes nonfuzzy; whereas in case of fuzzy subset, M is the interval [0,1].

The most fundamental question that one confronts is the basis of estimation of the truth values or membership grades of the pattern samples to each pattern class.

One of the major problems of using fuzzy set theoretic approaches to the pattern classification problem is the nonavailability of adequate tools for evaluation of membership function of a sample to a class of fuzzy patterns. Although a number of reports have come out suggesting various schemes of evaluation of membership grade, the major hindrance lies in constructing a homomorphism from a qualitative preference system to a quantitative preference system, which can always be corrupted by subjective considerations.

Techniques of estimation of parameters of membership function can be broadly grouped as

1. The techniques employing mathematical expressions like S and π curves.
2. The expressions involving similarity and distance measure.
3. Techniques using the statistical information.
4. Rank ordering of prototypes according to compatibility of the features in the set.
5. Scaling the membership function using representation and uniqueness theorem on an interval scale.
6. Estimation of membership degrees of an object using relative preference method by Saaty [16]. This is applicable in situations where the necessary initial information of the pairwise comparison is available from experts.

Some of the techniques suggest estimation of membership function by addition of several fuzzy variables with known membership grades by Dishkant [5]. The essence of each of the above approaches lies in the fact that most of these methods do not require large numbers of prototypes with known statistics for estimation of the membership function. Moreover the estimation of membership function should be so chosen that it is consistent with the specification of the data set.

Keeping in view the above, we now discuss a fuzzy supervised learning algorithm followed by unsupervised fuzzy classification techniques.

9.5 Fuzzy Supervised Learning Algorithm

Let the n-dimensional measurement vector of the kth prototype $x^k = x_1^k, \ldots, x_n^k$ be known. The membership grade of x to the fuzzy class c_i will be given as

$$\mu_{c_i}(X) = \left[1 + D(x^k, \overset{\frown}{X}_i)\right]^{-1} \tag{9.8}$$

where $\overset{\frown}{X}_i$ is a core pattern sample whose initial membership grade to the fuzzy class c_i, $i=\{1,\ldots, l\}$ is assumed to be unity and $D(X^k, \overset{\frown}{X}_i)$ represents the distance between the kth prototype and the core sample $\overset{\frown}{X}_{ii}$ of the ith fuzzy class. Henceforth, we will use the symbol $\in_f$ to indicate the belongingness of a sample to a fuzzy class of pattern. Here we assume a city block distance measure of the form:

$$D(X^k, \overset{\frown}{X}_i) = \frac{1}{m}\sum_{j=1}^{m} \alpha_{ij}\left|X_j^k - \overset{\frown}{X}_i\right| \tag{9.9}$$

where $|X_j^k - \overset{\frown}{X}_{ij}|$ yields the distance measure between the kth prototype X^k and the core sample $\overset{\frown}{X}_i$ of the ith class of pattern in the jth dimension.

As may be noted α_{ij} represents a $(n \times m)$ weight matrix where, for each class i, α_{ij} represents weight in jth dimension. The α_{ij}'s are the unknown parameters to be estimated during learning phase when finite large number of pattern samples are fed to the machine.

The estimation of the weight vector α_{ij} can be done using x prototypes or training samples during the supervised learning process. Once the weight vectors α_{ij} are estimated for all the fuzzy pattern classes, an unknown pattern x is classified to one of the m classes according to the principle of maximum membership.

The weight vectors α_{ij} are iteratively modified when a sequence of training samples are presented, such that

$$D(X^k, \overset{\frown}{X}) = \frac{1}{m}\sum_{j=1}^{m} \alpha_{ij}\left|X_j^k - \overset{\frown}{X}_{ij}\right| \le T_1 \quad \text{when } X^k \in C_i \tag{9.10}$$

$$D(X^k, \overset{\frown}{X}) = \frac{1}{m}\sum_{j=1}^{m} \alpha_{ij}\left|X_j^k - \overset{\frown}{X}_{ij}\right| > T_2 \quad \text{when } X^k \in C_i \tag{9.11}$$

Let α_{ij}^k be the weight vector after the kth labeled sample is presented to the trainer and let $\delta^k d_{ij}$ be the increment in distance when the kth pattern is presented. The conditions of the iterative modification of the weight vector α_{ij} are shown below.

Case 1: If $D(X^k, \overset{\frown}{X}_i) > T_i$ and $X^k \in c_i$, then the measure $d_{ij} = |X_{ij}^k - \overset{\frown}{X}_{ij}|$ and repeat the procedure below for all j.

a. If $\alpha_{ij}^{k-1} d_{ij} > t_{ij}$ then modify α_{ij} such that

$$\alpha_{ij}^k \leftarrow \alpha_{ij}^{k-1} - \rho \cdot \delta \cdot d_{ij}$$

where $\delta^k d_{ij} = d_{ij} - t_{ij}$ and ρ is a constant.

b. Else if $\alpha_{ij}^{k-1} d_{ij} < t_{ij}$ then do not change α_{ij}

Case 2: If $D(x^k, \hat{x}_i) < T_i$ and $x^k \in c_i$
then α_{ij} is not modified.

Case 3: If $D(x^k, \hat{x}_i) < T_i'$ and $x^k \in c_l$ $(l \neq i)$.
then measure, $d_{ij} = |X_{ij}^k - \hat{X}_{ij}|$ for every j and repeat the steps below for all j.

a. If $\alpha_{ij}^{k-1} d_{ij} < t_{ij}'$, then modify α_{ij} such that

$$\alpha_{ij}^k \leftarrow \alpha_{ij}^{k-1} + \rho \cdot \delta \cdot d_{ij}$$

where $\delta^k d_{ij} = t_{ij}' > t_{ij}$ and ρ is a constant

b. Else if $\alpha_{ij}^{k-1} d_{ij} > t_{ij}'$, do not modify α_{ij}

Case 4: If $D(x^k, \hat{x}_i) > T_i'$ and $x^k \in c_l$ $(l \neq i)$
then α_{ij} is not modified.

It may be noted here that in addition to two different threshold values T_i and T_i' which influence the trajectory of the weight vectors, a threshold band $(-t_{ij}, t_{ij})$ has been chosen for each dimension j and the weight vector is modified in a certain dimension j if $\alpha_{ij}{}^* d_{ij}$ moves beyond the dimensional threshold band.

In the light of the above discussion the weight vectors are iteratively modified during the supervised learning process at each step.

9.6 Fuzzy Partition

The concept of fuzzy partition is essentially hidden in the basic definitions of union and intersection of two fuzzy subsets as introduced by Zadeh [21] and discussed in Chapter 1. The mathematical systems involving fuzzy clusters are well documented in Bezdek [2].

The techniques of unsupervised pattern classification or cluster analysis encompasses a broad spectrum of methods whereby a data set x is partitioned in "n" subsets or clusters $\{x_i, \ldots, x_n\}$ which are pair wise disjoint and reproduce X via union. However, there is one inherent problem in all these methodologies arising from the underlying axiom that each point in X_i is unequivocally grouped with other members in the same cluster X_i and there is no way to characterize each point's similarity to other members of set X. This problem can be overcome by associating a membership value in the interval [0,1] to each sample in every cluster such that each sample shares a similarity with each cluster; membership values near 0 signify small similarity between the sample and the cluster whereas membership values near 1 designate a high degree of similarity.

The above concept leads to the foundation of fuzzy clustering, where a member can have partial membership grades in several fuzzy clusters. This is the distinction between the hard and fuzzy c-partitions of a data set X.

9.6.1 Pattern Classification Using a Fuzzy Similarity Measure

In this section we will present an algorithm based on hierarchical random partitioning of fuzzy classes of pattern. This method makes use of the similarity relations and fuzzy clustering.

The problem of partitioning a fuzzy set by similarity relations in terms of ordinary equivalence relations has been considered by several researchers.

A thorough investigation on the fuzzy clustering techniques utilizing optimization of objective functions can be found in Bezdek [2]. In case of hierarchical clustering methods, the computational and storage complexities become prohibitively large for implementation in many practical situations. A method has been presented here that describes hierarchical clustering using random partitioning method. A fuzzy similitude, defined as the limiting value of an n-step fuzzy relation, similar to the one defined in Tamura et al. [18], has been used for clustering the sample points inside the randomly generated partitions at each level of the hierarchical structure.

9.6.2 Fuzzy Similitude and Partitioning

Pattern classification using subjective information based on suitable similarity measure has been presented here. The method is based on the procedure of extracting a path between two patterns and can be applied to the problem of fuzzy classification. This method utilizes a λ threshold relation for pattern clustering as in Tamura et al. [18].

Let $X=\{x_1, \ldots, x_n\}$ be a finite set of patterns. To characterize the fuzzy relation between this patterns, a fuzzy similarity function can be defined. We start with the following definitions:

$$f_{\mathrm{R}}(x,y) \in [0,1], \quad \text{for all } x, y \in X$$

The fuzzy relations between two pattern elements x and y can be viewed as an n-ary fuzzy relation defined as

$$f_{\mathrm{R}}^{n}(x,y) = \max_{x_1,\ldots,x_n \in X} \min\left[f_{\mathrm{R}}(x,x_1), f_{\mathrm{R}}(x_1,x_2)\ldots, f_{\mathrm{R}}(x_{n-1},y)\right] \quad \text{for } n = 2,3,\ldots \tag{9.12}$$

Also

$$f_{\mathrm{R}}^{n+1}(x,y) = \max_{x_1,\ldots,x_n \in X} \min\left[f_{\mathrm{R}}(x,x_1)x_1,\ldots, x_n \in X\ldots f_{\mathrm{R}}(x_{n-1},x_n), f_{R}(x_{n},y)\right]. \tag{9.13}$$

It can be seen that

$$f_R^{n+1}(x,y) \geq f_R^n(x,y) \tag{9.14}$$

Thus for all $x, y \in X$ and all $n \geq 1$,

$$0 \leq f_R(x,y) \ldots \leq f_R^n(x,y) \leq f_R^{n+1}(x,y) \leq 1 \tag{9.15}$$

The limiting value of $f_R^n(x,y)$ exists by the monotone convergence, that is, for every $\in > 0$, there exists an $n < N$ such that

$$\left| f_R^n(x,y) - \widehat{f_R^n}(x,y) \right| < \in \quad \text{for } n > N \tag{9.16}$$

where $\widehat{f_R^n}(x,y)$ is the limiting value of $f_R^n(x,y)$ for $n \to \alpha$. The relations as defined above may be reflexive, symmetric, and transitive. Amongst various types of reflexivities, we define here weak reflexivity.

Definition 9.1
R is weakly reflexive if $f_R(x, x) \geq f_R(x, y)$ for all y in U.

Definition 9.2
A fuzzy relation is symmetric if $f_R(x,y) = f_R(y,x)$ for all x, y in U.

Definition 9.3
A fuzzy relation R is transitive if $R \geq R \circ R$ where $R \circ R$ denoted by R^2, is characterized by

$$f_{R \circ R}(x,y) = \max_{u \in U} \min \left(f_R(x,u), f_R(u,y) \right) \tag{9.17}$$

It can be noted that ordinary reflexivity where $f_R(x, x) = 1$ is a special case of weak reflexivity.

The above notion of transitivity plays a great role in partitioning and clustering. If a fuzzy matrix (relation) is transitive then $R \geq R^2 \geq R^3 \cdots \geq R^n$

If $R^K = R^{K+1}$ for some positive integer k then R is called convergent. It can be proved that if R is convergent, then R is transitive. The proof follows from the monotonicity of the transitive relation.

Although in most of the pattern classification problems the fuzzy relation between two patterns becomes a symmetric one, there are situations where symmetric affinity is observed amongst the data set. An asymmetric measure is thus necessary to characterize such situations. Starting with a generalized asymmetric fuzzy relation R, we can identify the symmetric Kernel R_K from R where R_K is a symmetric relation.

Definition 9.4
The symmetric kernel R_K of the fuzzy asymmetric matrix R is given as

$$R_K = R \cap R' \tag{9.18}$$

where R' is transpose of R.

We now state the following proposition.

Proposition 9.1
If R is transitive then R_K is also transitive.

Proof. From the definition of transitivity, R is transitive if $R \geq R \circ R$. Also from definition of R_K

$$R_K \leq R \quad \text{and also} \quad R_K \leq R'$$

Thus it follows that

$$(R \cap R') \circ (R \cap R') \leq R \circ R \leq$$

and

$$(R \cap R') \circ (R \cap R') \leq R'R' \leq R'$$

This is true since R' is transitive if R is transitive.

Therefore

$$R_K \circ R_K \leq R \leq R'$$

$$R_K \circ R_K \leq R_K$$

Hence R_K is transitive.

A fuzzy relation which is reflexive, symmetric, and transitive is a fuzzy similitude (equivalence relation). The fuzzy equivalence relation as defined above induces partitioning on a fuzzy pattern space. The pattern x and y can be clustered in one partition if we can define a threshold relation $R_T(x, y)$ such that

$$f_{R_T}(x, y) \geq T \tag{9.19}$$

where T is an appropriately defined threshold.

We now state the following theorem [18].

Theorem 9.1
The threshold relation is an equivalence relation.

Proof

1. xR_Tx is true for any type of reflexivity
2. xR_Ty iff yR_Tx, since $f_{R_T}(x,y) = f_{R_T}(y,x)$
3. xR_Ty, $yR_Tz \Rightarrow xR_Tz$ since $f_{R_T}(x,z) = \max \min\left[f_{R_T}(x,y), f_{R_T}(y,z)\right]$

Thus a threshold relation induces a partition with the threshold T.

Using the above threshold similarity relation we can now design the fuzzy hierarchical classifier. As has been mentioned, the pattern samples at each level are randomly partitioned and in each partition the pattern samples are clustered using fuzzy threshold relation.

The choice of the representative core sample from each cluster is guided by the principle that core sample representing the kth cluster should be such that it will have the greatest neighborhood mass function. The definitions are as follows:

Definition 9.5
A λ-neighborhood of the mass pattern $x \in X$ is defined as

$$V(x,\lambda) = \left[y \in X \middle| f_R(x,y) > \lambda\right] \tag{9.20}$$

where $f_R(x,y)$ is the similarity relation between patterns x and y.

Definition 9.6
The neighborhood mass function of a sample $x \in X$ is defined as

$$M(x,\lambda) = \text{card}\, y \middle| f_R(x,y) > \lambda \tag{9.21}$$

λ is a user defined value.

The neighborhood mass function of each of the sample points in a cluster is determined and the sample with maximum mass function is chosen as the core sample representing the pattern cluster.

The selection of the threshold T for fuzzy partitioning at each level is guided by the affinity coefficient existing in the fuzzy data set.

We define a fuzzy dispersion coefficient in each cluster which yields a measure of the maximum dispersion of the sample point with reference to the core in a particular cluster.

Definition 9.7
The dispersion coefficient of the ith pattern cluster is

$$\rho_L = \max\left[d\left(\hat{x}_l, x\right)\right]$$

$$\forall x \in \mathrm{CL}^{i}(l) \tag{9.22}$$

where $d(\overset{\frown}{x}_l, x)$ is a distance measure from the core sample $\overset{\frown}{x}$ to the sample x of the lth cluster. Next an affinity coefficient $\alpha(x)$ of the lth cluster can be given as

$$\alpha(x) = 1^{1}{}_{\frac{1}{1+\rho(x_l)}} \tag{9.23}$$

It can be observed that the affinity coefficient yields a minimum similarity measure between the core sample and the other samples farthest from the core sample of the cluster.

Thus for each core pattern sample $\overset{\frown}{x}_1^i, \overset{\frown}{x}_2^i, \ldots, \overset{\frown}{x}_l^i$ in the ith partition, at a certain level, we can associate a set of corresponding affinity coefficients $\alpha_1, \ldots, \alpha_l$.

The threshold T for fuzzy clustering in the said partition at a certain level can be adjusted such that $T \geq \min(\alpha_1, \ldots, \alpha)$. Thus the threshold T at each level of hierarchical clustering is adaptively changed in accordance to the affinity coefficients associated with each care sample of the fuzzy clusters at the previous level.

The algorithm terminates when all the pattern samples have merged to form the final clusters. The algorithm of the hierarchical clustering is given below.

Procedure: Fuzzy hierarchical clustering

Step 0: Set the level $k=0$; choose initial value of threshold T; set $R^k = \{x_1, \ldots, x_n\}$.

Step 1: Partition n-sample points $\{x_1, \ldots, x_n\}$ in m partitions $p_1, \ldots, p_m$ in a random fashion.

Step 2: Group the pattern points in each partition p, $i=1, \ldots, m$ to form clusters $\mathrm{CL}^i(j)$, $j=1, \ldots, l$, using fuzzy similarity threshold relation.

Step 3: Extract the core points $\overset{\frown}{x}_1^i, \ldots, \overset{\frown}{x}_l^i$ from each of the clusters $\mathrm{CL}^i(j)$ within each partition P_{pi}, $i=1, \ldots, m$ where

$$\overset{\frown}{x}_j^i = \max\left[M(x,\delta)\right]; \quad M(x,\delta) \text{is the mass}$$

$$x \in \mathrm{CL}^i(j)$$

function of the sample $x \in \mathrm{CL}^i(j)$.

Step 4: Determine the fuzzy affinity coefficient

$$\alpha_j,\ j=1,\ldots,l \quad \text{for each cluster } CL^i(j)$$

The threshold T for fuzzy clustering in the said partition at a certain level can be adjusted such that $T \geq \min(\alpha_1, \ldots, \alpha_l)$. Thus the threshold T

at each level of hierarchical clustering is adaptively changed in accordance to the affinity coefficients associated with each core sample of the fuzzy clusters at the previous level.

Step 5: $K=k+1$; consider each core point from each cluster at kth level to form a set $R^K = \{x^i(j) \mid j = 1,...,\lambda$ and $i = 1,2,3\ldots,m\}$. Adjust threshold T such that $T \geq \min [\alpha_1, \ldots, \alpha_l] \forall\alpha_i \in R^K$

$$\forall\alpha_i \in R^K$$

Step 6: Determine card $[R^k]$ where card $[R^k]$ is the number of representative core samples in level K. If card $[R^k] \leq C_m$, the number of pattern classes, then go to step 7. Else go to step 2.

Step 7: The clustering procedure is over.

9.7 Fuzzy Unsupervised Pattern Classification

The problem of fuzzy clustering can formally be presented as below.

Let $X=\{x_1, \ldots, x_n\}$ be the finite set of n pattern samples to be partitioned in C $(2 \leq C \leq n)$ pattern classes $\{u_1, \ldots, u_c\}$, where the ith partition is represented by u_i. Considering U_{kj} to be the membership grade of pattern x_j to class and $U=[u_{kj}]$ is a real C^*n matrix the following conditions yields fuzzy c-partition of the data set.

$$u_{kj} \in \{0,1\}$$

$$\sum_j u_{kj} > 0, \quad \text{for each } k = 1,\ldots, C \tag{9.24}$$

and $\Sigma_j u_{kj}=1$.

Thus the set of the fuzzy c-partitions of x can be given as

$$M_{fc} = \left\{ u \middle| u_{kj} \in [0,1]; \sum_j u_{kj} > 0, \sum_k u_{kj} = 1 \right\} \tag{9.25}$$

The fuzzy partition enriches the conventional hard partitioning model. This implies that hard c-partition can be obtained from fuzzy clustering.

As in the case of hard partitioning, several criteria have been proposed in the literature for identifying optimal fuzzy c-partitions in the data set X.

Fuzzy Isodata clustering algorithms have been suggested for fuzzy partitioning of a finite data set and have been implemented by minimization of least square objective functionals.

The often used least squared objective functional is

$$J_m(u,v) = \sum_{j=1}^{N} \sum_{k=1}^{C} (u_{kj})^m \|x_j - v_k\|_A^2 \quad (9.26)$$

where

u is a fuzzy c-partition of data set x

v indicates vector of cluster centers

m is an weighting exponent $1 \le m \le \alpha$.

$$\text{The } A\text{-norm } \|x_j - v_k\|_A^2 = (x_j - v_k)^T A (x_j - v_k) \quad (9.27)$$

where A is a positive definite weight matrix, indicates squared A-distance form x_j to each cluster center v_k, If $m=1$, the algorithm behaves more like the classical isodata process whereas for large values of m the extremal partitions tend to be very fuzzy, regardless of whether clusters are present in data sample.

A number of measures have been proposed for assigning scalar measures of partition fuzziness. Ruspini [15] introduced a partition entropy concept

$$H(k;u) = -\frac{1}{n\log(k)} \sum_{i=1}^{k} \sum_{x \in X} u_i(x) \log_a (u_i(x)) \quad (9.28)$$

Dunn [9] suggested a different coefficient that measures the overlap in the membership of the extremal partitions of u.

$$F(k,u) = \frac{1}{n} \sum_{i=1}^{k} \sum_{x \in X} u^2 \quad (9.29)$$

Apart from the above fuzzy validity functional, a number of other functionals have been reported in the literature.

The mean and variance of the above two validity functional have been estimated under certain statistical assumption on the distribution of fuzzy c-partition space.

The fuzzy clustering procedures have been related to the estimation of mixture distribution of statistical pattern classification techniques using the Euclidean metric. Some of the important application of fuzzy c-means algorithms is as follows:

1. The fuzzy partitioning algorithms yield better results than hard clusters if the substructure is imprecise.
2. When data set has distinctive substructure, fuzzy and hard cluster yield the same feature selection results.

In an imprecisely defined medical diagnosis problem involving six classes of diseases with 11 measured symptoms, a lower expected error rate by using fuzzy 1-NP classifier over conventional K-NN classification results using the same data set. The fuzzy nearest prototype classification algorithm operates by minimizing the distances from each object x_j in x from one of the c-prototypes $v_1, \ldots, v_c$ which are ordinarily the geometric centroid of each class and characterize the perfect sample of the classes.

Fuzzy c-means have found application in shape description, where a boundary of a planar shape can be approximated with a set of approximating line segments $\{L_i, 1 \leq i \leq C\}$. The optimal set of L_i can be generated using several approaches like split and merge algorithm and fuzzy c-partition approach. A weighted least squared error function of three sets of variables, namely, fuzzy c-partitions, nuclei vectors, and direction vectors have been minimized to identify the fuzzy clusters in a data set of ordered boundary coordinates of a planar shape and a set of c line segments for a good least square fit to the data.

9.8 Summary

We have presented the fundamentals of pattern classification that includes various supervised and unsupervised classification schemes. The limitations of statistical classifiers have been identified, followed by a discussion on how fuzzy techniques in pattern classification will be more appropriate in many applications. Fuzzy supervised classification, fuzzy clustering, and fuzzy similitude–based classifier have been elaborated.

References

1. Bezdek, J.C., A convergence theorem for the fuzzy c-means clustering algorithms, *IEEE Transactions on PAMI*, 2, 1–8, 1980.
2. Bezdek, J.C., *Pattern Recognition with Fuzzy Objective Function Algorithms*, Plenum Press, New York, 1981.

3. Bezdek, J.C., Ehrlich, R., and Full, W., FCM: The fuzzy clustering algorithm, *Computers and Geosciences*, 10(2–3), 191–203, 1984.
4. Bishop, C.M. (Ed.), *Neural Networks and Machine Learning*, 1997 NATO Advanced Study Institute, Springer, Cambridge, U.K., 1998.
5. Dishkant, H., About membership function estimation, *International Journal of Fuzzy Sets and Systems*, 5(2), 141–147, 1981.
6. Dubes, R. and Jain, A.K., Clustering techniques: The users dilemma, *Pattern Recognition*, 8, 247–260, 1976.
7. Duda, R.O., Hart, P.E., and Stork, D.G., *Pattern Classification*, 2nd ed., John Wiley & Sons, New York, 2001.
8. Duda R.O. and Hart, P.E., *Pattern Classification and Scene Analysis*, Wiley, New York, 1973.
9. Dunn, J.C., Indices of partion fuzziness and the detection of clusters in large data sets. In Gupta, M.M., (Editor), *Fuzzy Automata and Decision Process*, Elsevier, New York, 1976.
10. Friedman, M. and Kandel, A. *Introduction to Pattern Recognition, Statistical, Structural, Neural and Fuzzy Logic Approaches*, World Scientific, Singapore, 1999.
11. Fu, K.S., *Syntactic Pattern Recognition and Application*, Prentice-Hall, Englewood Cliffs, NJ, 1982.
12. Fukunaga, K., *Introduction to Statistical Pattern Recognition*, 2nd ed., Academic Press, New York, 1990.
13. Kohonen, T., *Self Organising Maps*, Springer, Berlin, 1997.
14. Majumdar, A.K. and Ray, A.K., Inference of fuzzy regular pattern grammar, *Pattern Recognition Letters*, 2, 27–32, Oct. 1983.
15. Ruspini, H., A new approach to clustering, *Inform and Control*, 15(1), 22–32, 1969.
16. Saaty, T.L., Exploring the interface between hierarchies, Multiple objects and Fuzzy sets, *Fuzzy Sets and Systems*, 1, 57–68, 1978.
17. Shawe-Taylor, J. and Cristianini, N., *Kernel Methods for Pattern Analysis*, Cambridge University Press, Cambridge, U.K., 2004.
18. Tamura, S., Higuchi, S., and Tanaka, K., Pattern classification based on fuzzy relation, *IEEE Transactions on SMC*, SMC-1, 61–66, 1971.
19. Theodoridis, S. and Koutroumbas, K., *Pattern Recognition*, Academic Press, Amsterdam, the Netherlands, 1999.
20. Tou, J.T. and Gonzalez, R.C., *Pattern Recognition Principles*, Addison-Wesley Reading, MA, pp. 119–123, 1974.
21. Zadeh, L.A., Fuzzy sets, *Information and Control*, 8, 338, 1965.

10

Application of Fuzzy Set Theory in Remote Sensing

10.1 Introduction

Remote sensing is the science of marking inferences about objects from measurements made at a distance. Remote sensing of earth resources using space-borne sensors is accepted as a useful tool by the application scientists for generating inputs to the natural resources management system. Undoubtedly, it is the fastest and cheapest means of obtaining information of a vast area of the earth's surface. The data are obtained using satellite-borne sensors available in the optical and nonoptical ranges of electromagnetic spectrum.

The fingerprints or signatures of the terrestrial features are stored and used for identification of land cover for mapping and resources planning in the fields of forestry, hydrology, geology, oceanography, and agriculture. The land cover is one of the most fundamental geographic variables for utilization of resources in these fields. The accuracy and reliability of this mapping can go a long way in the natural resources management of a nation. The advantages of satellite-borne remote sensing are synoptic coverage, repetitive coverage, and multiple sensor output useful in distinguishing the objects.

10.2 Why Fuzzy Techniques in Remote Sensing?

The automated identification of land surface features from remotely sensed (RS) data is a complex task and has assumed paramount significance in recent times. The interaction of electromagnetic radiation with natural surfaces cannot be properly described due to the presence of a large number of variables and the uncertainties associated with the variables. It is thus difficult to appropriately quantify the surface reflectance. The other factors affecting the process of remote sensing are the nonlinearities of the sensors, satellite attitude at the time of viewing, earth rotational errors, sun error, atmospheric haze, and attenuation. In spite of correcting the data for most of these errors, imprecision still remains.

Moreover, the remote sensing image data is only a 2-D projection of 3-D natural world and some amount of imprecision is inherent in the images. Often, the boundaries and properties of the region are found to be vague and fuzzy in RS images. Uncertainties coupled with statistical randomness give rise to many errors in the identification of land surface which make it a challenging task. This necessitates the use of mathematical tools based on imprecision to model and process this inexactitude and associated uncertainties into the remote sensing frame work.

Fuzzy logic provides some useful concepts in overcoming the expressive inadequacy and permits pixels belonging to more than one category with graded membership. This varying degree of membership is also useful in resolving the mixed pixel to its component classes. Fuzzy techniques are employed in cases where the imprecision in the pattern description is interpreted from an intrinsic ambiguity or imprecise definition, which is not connected with the randomness of the environment. Apart from increasing the accuracy of classification, the area extent of the land cover class can be estimated more accurately by applying fuzzy methods.

Artificial neural networks (ANN) are mathematical models to emulate the biological neural networks found in the cerebral cortex of the human brain. Multilayer perceptron (MLP) with back-propagation learning and Kohonen's neural network have been widely applied in the recent past for multispectral RS image classification.

A new class of neural networks known as fuzzy neural networks is becoming widely popular due to their combined advantages of fuzzy logic and neural networks. Though the fuzzy neural networks have been applied to a few scientific problems, these have not been extensively applied to the classification of RS images.

10.3 About the Remotely Sensed Data

The satellites meant for earth observation are, in general, polar-orbiting, sun-synchronous satellites. The orbital plane of the satellites encompasses the poles and they cross the equator at the same local time at a specific place on earth every day. Each satellite is on a fixed path and calendar which is inflexible. The sensors on the satellite platforms collect the electromagnetic energy from the earth due to radiation, scattering and reflection from the major source, that is, the Sun.

Each sensor is designed to record a specific band of electromagnetic radiation (EMR) spectrum with certain clarity of ground detail and periodic repetition. The resolution of the RS data refers to the effectiveness and utilization of RS data. Spectral resolution refers to the number of wavelength intervals (bands) in which the data is collected. A narrowband or a specific wavelength interval is chosen based on the characteristics of the object under scrutiny, atmospheric absorption, and the background objects.

The spatial resolution decides the size of the feature to be identified on earth. It is a measure of the smallest angular or linear separation between two objects that can be resolved by the sensor. In an aerial image, it is the dimension of the ground projected in the instantaneous field of view. The dimensions on ground depicted by each pixel could be $79\,\text{m}\times 79\,\text{m}$ or $36\,\text{m}\times 36\,\text{m}$ depending upon the sensor and satellite. Enhancement of resolution results in increase of data. For example, spatial resolution improvement by a factor of two increases the volume of data fourfold.

The bands are chosen after a rigorous study of the parameters involved and the end use of the data. A large bandwidth (0.4–0.7 μm) is sometimes used for obtaining panchromatic, that is, black-and-white image with greater spatial details. On the other hand, narrowband (0.63–0.69 μm) images are used for study of urban areas.

Temporal resolution is the ability of the sensor to provide repetitive coverage of the same area at different times. The satellite orbital characteristics govern this ability. Repetitiveness and revisit capability are the two important variables which decide temporal resolution of data. These are essential to measure the temporal data, such as the change of crop cycles in agriculture. Radiometric resolution refers to the sensitivity of the sensor for the light flux. It also refers to the number of discriminable levels of each pixel.

10.4 Classification of Remotely Sensed Data

Classification of multispectral RS involves assignment of a RS pixel with different spectral bands to a proper pattern class. A pixel vector in a RS image

$x = \begin{bmatrix} x_1 \\ \cdot \\ x_n \end{bmatrix}$ of dimension n is assigned to one of the m classes $\{C_1, \ldots, C_m\}$.

During the process of classification, the data is transformed into information by assigning individual pixel of a multispectral image to discrete and distinct categories.

In a supervised classification, identification and location of some of the land cover types such as urban, agriculture, wet land, and forest are known a priori. The knowledge of the type of cover, known as ground truth, could be obtained be field work, maps, aerial photography, experience of the human interpreters, or a combination of more than one of these methods. The samples of selected areas of land cover thus obtained are represented as labeled pixels that constitute the training set.

10.5 Fuzzy Sets in Remote Sensing Data Analysis

The data obtained through space-borne sensors is imprecise in nature due a number of processes affecting it. The reflectance, emission, and absorption of an object in the visible and invisible parts of the electromagnetic spectrum are the deciding factors which characterize the object in the image. A thorough knowledge of the mechanism is necessary to interpret the data thus generated. The emissivity of the surface depends upon wavelength, temperature, and geometry of the surface. The reflectance and/or scattering effect is a fraction of the intensity of the incident radiation. The electromagnetic properties of the substance thus play an important role in the manner in which they appear in the image.

Correlation of wavelength-dependent reflectivity of a target to its spectral signature is obtained in the form of ground truth. Some of the anomalies observed in the signature or the fingerprint of the class of objects are that two object classes can sometimes give identical signatures such as land and soil, urban land and bare soil. Pixel sizes larger than the object width can result in mixed pixels. The spectral mixture of two land use categories can give rise to different spectra, as in the example of bare soil and vine leaves. A single pixel containing a mixture of two or three classes can be wrongly classified to any one particular class leading to error in the measurements in terms of area coverage.

Spectral overlap of land covers classes and pixel forming at the boundary of regions lead to erroneous classification results. Conventional methods are found to have expressive inadequacy while assigning each pixel to a single class in the training stage itself [27]. Mixtures of spectral reflectance inside the pixel make such categorization false, leading to inaccuracies in classified output. Pixel sizes larger than the width of the observed object make it problematic to assign it uniquely to a particular category.

An accurate mathematical model from the first principles cannot be derived due to the uncertainties associated with the atmospheric phenomenon. Generalization of such factors is carried out experimentally to aid the analysis of the RS image. Accuracy in measurement of these errors and reversal of the processes limit the final accuracy of the processed output.

The mapping of the land cover for earth resources monitoring has not achieved the operational status due to the fact that quite often, the classification reliability is not satisfactory and confidence measures are not available.

Automated methods of information extraction are favored in remote sensing images due to the enormous amount of data received from satellite. The problem is compounded by the multiband, multi-temporal data, specific to a particular area. Automatic and machine-oriented methods of pattern recognition have been widely used in the characterization and abstraction of such data. Improvements in sensor and satellite, resulting in better resolution and better revisit capacity demand faster and more accurate methods to make the best use of the available data.

The inherent fuzziness associated with the RS images advocate the need of inputs, outputs, and processing in the fuzzy domain. The advantages envisaged in such a scenario are better accuracy, proper representation of input values in fuzzy domain, and output in the form of pixel memberships indicating their partial belongingness to a category.

10.6 Background Work in Neuro Fuzzy Computing in Remote Sensing

In the past, MLP classifiers with back-propagation learning (generally referred as ANN in literature) have been extensively used as land cover classification by researchers in remote sensing.

Hepner [16] showed that an ANN can be trained to do land cover classification of satellite imagery using selected sites representative of each class in a manner similar to conventional supervised classification. He compared the results to MLC method. ANN application for classification was analyzed in the light of compact training sets and it was found that a single training site per class was comparable to four training sets per class in conventional classifier. Benediktsson et al. [1] conducted extensive experiments with ANN in multi-source data classification and compared the output accuracy with four different statistical methods. The goal was to increase the influence of more reliable data source in decision making and vice versa. A global membership function was defined based on the individual reliability. It is required that each source should have a reliability measure. Three measures, namely, weighted average repeatability; overall classification accuracy, and equivocation were identified. It was proposed that in the repeatability, if the classes are high, it is considered to be a more reliable source. Similarly, if the classification accuracy is high, it is considered to be a more reliable source. Equivocation is the quality that the data class strongly indicates information class on the map of ground truth. Gray codes were used at the input. The output was to be classified into one of the 10 classes. In addition to four band multispectral data, the authors used data on elevation, slope, and aspect ratio. The output accuracy was shown to be better than with statistical methods like minimum Euclidean, maximum likelihood (Gaussian), MLC with reliability, and M-distance measure with Gaussian. Lee et al. [20] used the ANN for texture-based cloud classification using nine textural features in the input. Accuracy was found to be better than 90% on each type of cloud classification. Earlier schemes for such an application were linear, parametric predictions. The comparison was made between nonparametric k-Nearest Neighbor approach and linear step-wise discriminant analysis procedure.

Higher accuracy was obtained with only 20% of the database as the training data, compared to 67% data used in the linear approach. The K-NN approach showed 85% accuracy and linear discriminant approach showed 84%. However the ANN method was found to be very slow, taking 120,000 iterations to converge to the reported accuracy.

Bishof et al. [4] used coarse coding at the input to discriminate similar classes. This involves use of 13 nodes per hand in the input layer. In addition, 7×7 pixels of texture was also fed to as many input nodes to increase the classification accuracy. It was suggested by the authors that by studying the weight visualization (W-V) curves, it was possible to eliminate one of the two highly correlated bands without sacrificing the accuracy of classification. A post classification smoothing was performed by contextual neighborhood consideration and the confidence level obtained during the classification. Due to post classification smoothing, an improvement of accuracy was demonstrated.

Heermann and Khazenie [15] used the 24 bit pattern of the three band data as input for classification of RS. A preprocessing method known as picking and packing to eliminate conflicting pixels as well as duplicate training samples was advocated by the authors. A detailed comparison of training time and classification time by conventional methods and ANN was presented. The slow training aspect was highlighted and an adaptive learning method and parallel coarse grain microprocessor machines were suggested as remedies.

As the number of hidden layer nodes increases the complexity, it was suggested that the least significant neurons in the hidden layer can be removed one by one to have a compact and optimized neural network. However, specific basis to accomplish the removal was not suggested [12].

An excellent review on the various aspects of MLP with back-propagation learning networks for classification of RS multispectral imagery can be found in Paola and Schowengerdt [28]. They have summed up the conclusions of various authors and analyzed in detail the various network parameters. In the case of input feature selection, it was observed that de-correlating the data in terms of its principal components can give much better accuracy and faster training of the network. The inputs with only mean vectors do not give any improvement in training accuracy. Textural features like second moment and entropy increased the discrimination capability.

The outputs can be assumed to give a level of confidence measure though rigorous proofs are not available to substantiate this proposal. There is a second proposal of assuming that the output directly gives the mixture of classes. While selecting the output class, the node with the highest output has been selected or two nodes with considerably higher membership were selected to resolve the mixed pixels. If all the outputs were below certain threshold value, the input is declared not to belong to any class.

While analyzing the network structure, the authors have observed that more number of hidden layer nodes gave more flexibility partitioning the decision space. However, no specific guidelines could be worked out with the available literature. In a peculiar and rare case, middle layer with four blocks of seven units in each block were used. Algorithms were by and large the same except that in a few cases, learning rate and momentum constants were adaptively reduced as the training progressed. The criteria for terminating the training could be based on the number of iterations, mean squared error or saturation of output to 0.9 and 0.1 for signifying the levels of 1.0 and 0.0, respectively.

The advantages of the ANN method are better accuracy, smaller training sets resulting in less labor in obtaining ground truths, ability draw complex decision boundaries, and flexibility of adding a number of inputs. The disadvantages, as highlighted by the authors are lengthy training time, loss of interpretability and nonstandard selection of parameters. It is pertinent to mention that in some of the cases, the conventional methods were found to be more accurate than the ANN methods.

Foody et al. [12] studied the effect of training set size and composition on an ANN. It was reported that the technique is more appropriate for classification when the training set is limited in size. The results were shown to be more accurate. They have concluded that training set size and composition are important variables which affect the classification accuracy and time while using the neural networks for classification.

Foody and Arora [13] evaluated the factors affecting the accuracy of the classification in ANN. These factors are network architecture, training sets size, discriminating variables, and testing data characteristics. It was found that the variables other than the network architecture had a significant effect on the performance.

Foody et al. [14] analyzed the method of unmixing composition of the pixels using ANN classifier. Though the complexity of the network and the intensity of training increased, the results of the case study showed that accurate estimates of the sub-pixel components could be made. A novel method of training the network on mixed pixels, eliminating the requirements for end-member spectra was followed. The area extent of classes could also be estimated more accurately.

Self-organizing feature map has been applied for clustering and subsequent classification [19]. In a modified vector quantization scheme, the feature map is trained first and for each input vector, a codebook vector is selected. The identification land features could be done in this manner. An aerial image with a very higher resolution was used for providing the scheme.

Neural network in its basic form has got attractive properties of learning from examples, nonparametric approach, and general applicability to mapping function type of problems. However the number of choices and parameters required to set up a training is overwhelming. Sometimes, a wrong set of initial weight may never converge to a required accuracy. There is no standard method of selecting the number of hidden layers and the neurons thereof. Still, a number of parameters are selected by extensive trials. Thus the method becomes problem specific and data dependent, though it has an universal applicability. An exclusive set of architecture and parameters for solving a particular problem may not work well for another problem.

10.7 Background Work on Fuzzy Sets in Remote Sensing

Fuzzy-based classification of multispectral data was compared with texture-based classification by Palubinskas et al. [29]. Fuzzy measure was obtained from conventional maximum likelihood classifier or M-distance classifier. The strength of class membership thus obtained was applied in fuzzy post classification operation.

Fuzzy classification of multispectral RS data was applied to estimate sub-pixel component [2,5,24,31]. A linear model and a regression model using fuzzy membership functions derived from the fuzzy c-means algorithm [6] were employed to estimate sub-pixel land cover composition [24]. The results were noticeably better in linear mixture model. However, the authors investigated fuzzy method for an operational use of tropical forest extent estimation and found that the prediction was much closer (99%) to the actual area than the conventional classification methods.

Fuzzy land cover classification was found to be more accurate and realistic representation of land cover than conventional hard classification techniques [7]. Multiple and partial class memberships, fundamental to fuzzy techniques are well illustrated by this method. Fuzzy c-means algorithms were used and it was compared with the results obtained from discriminant analysis and ANN methods.

The coarse resolution of RS data is available at a lesser cost. For a larger area coverage, for example, forest cover mapping, such data are favored for their cost effectiveness and reduced quantum of data processing. Sub-pixel forest estimates could be estimated accurately from means of applying fuzzy methods. Based on a 79 m resolution data spatially degraded to a resolution of 1.2 km, Foody [14] showed that a degree of agreement of 75.2% was obtained between the predicted and actual class for a four class classification of forest using fuzzy c-means algorithm.

Wang [31] conducted studies on fuzzy supervised classification using a soft maximum likelihood classifier. The training was based on a previously determined land cover in a fuzzy representation. The mean and covariance matrix of a conventional maximum likelihood classifier was replaced with fuzzy mean and fuzzy covariance matrix which were calculated by modifying with fuzzy membership values, drawn from the training data set. The membership grades estimated were found to be in close agreement with the actual land cover. The important assumption of this work is that the fuzzy training data set is available a priori which is practically a tedious and expensive preposition. Such data is more often not available with adequate accuracy.

Maselli et al. [24] used the same soft maximum likelihood classifier suggested by Wang for carrying out quantitative investigations on the actual relationships existing between fuzzy membership grades obtained by calculations and the reference pixels. MLC with a priori probabilities derived from the frequency histograms were employed.

10.8 Segmentation of Remote Sensing Images

Match-based segmentation is used to extract the regions in the remote sensed images. The image in Figures 10.1 and 10.2 contains 500 MB of raw pixel information in the six bands (three in visible region and three in near-infrared region) in Western Washington State. The image was corrected for atmospheric and terrain distortions, and georeferenced. It contains large areas of conifer trees, deciduous trees, and forested areas close to water.

In these results, HSV (hue, saturation, and value) color model is used for region extraction.

The conversion of RGB to HSV is as follows [17,30]:

For $r,g,b \in [0, 1]$, then transformation gives $h,s,v \in [0, 1]$ as follows.

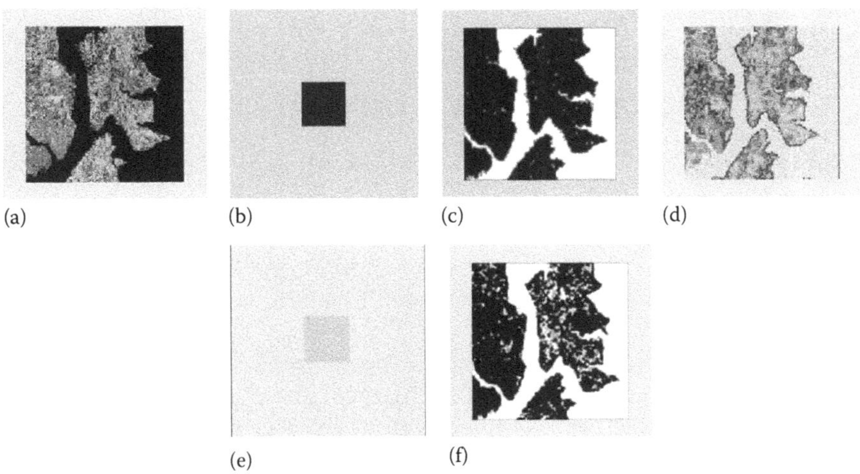

FIGURE 10.1
(a) **(See color insert following page 144.)** Remote sensed image, (b) query image, (c) region extracted (white region) using RGB model, (d) **(See color insert following page 144.)** image of (a) in HSV model, (e) query image as in (b) in HSV, and (f) region extracted using HSV model (white region).

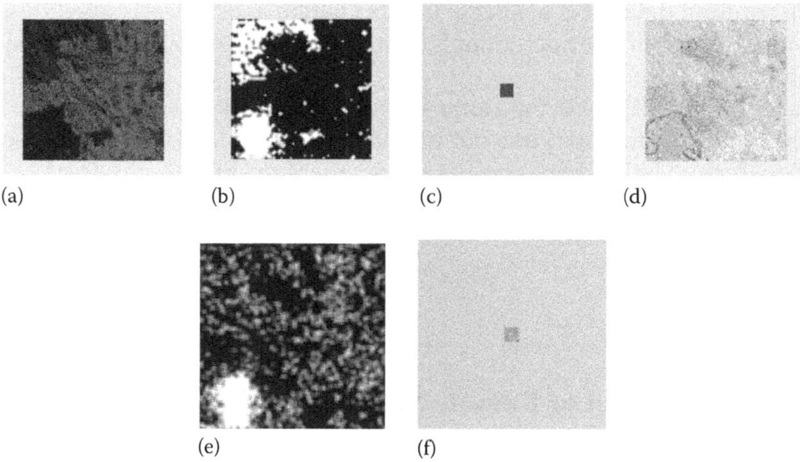

FIGURE 10.2
(a) **(See color insert following page 144.)** Remote sensed image, (b) **(See color insert following page 144.)** region-extracted image of (a) in RGB model, (c) **(See color insert following page 144.)** query image, (d) **(See color insert following page 144.)** image of (a) in HSV model, (e) **(See color insert following page 144.)** region extracted using HSV model, and (f) query image as in (c) in HSV model.

Intensity: $V = \max(r, g, b)$

Saturation: $S = [V - \min(r, g, b)]/V$

Let $r1 = (V - r)/[V - \min(r, g, b)]$, $g1 = (V - g)/[V - \min(r, g, b)]$
$b1 = (V - b)/[V - \min(r, g, b)]$

$$6h = \begin{cases} 5 + b1 & \text{if } r = \max(r, g, b) \text{ and } g = \min(r, g, b) \\ 1 - g1 & \text{if } r = \max(r, g, b) \text{ and } g \neq \min(r, g, b) \\ 1 + r1 & \text{if } g = \max(r, g, b) \text{ and } b = \min(r, g, b) \\ 3 - b1 & \text{if } g = \max(r, g, b) \text{ and } b \neq \min(r, g, b) \\ 3 + g1 & \text{if } b = \max(r, g, b) \text{ and } r = \min(r, g, b) \\ 5 - r1 & \text{otherwise} \end{cases}$$

The match-based segmentation algorithm remains the same as described in Chapter 6 when working on the RS images. The query image resembles any type of region, such as vegetated region, volcanic region, or other types of regions and the region that corresponds to those regions in the RS image is extracted. The procedure remains the same as described for color images; the only difference is that conversion from RGB to HSV, which is a user-oriented color model, is performed at the initial stage using MATLAB® 5.1 command.

In Figure 10.1, the query image is a part of the RS image. In HSV model, the region is extracted very clearly which is not the case for RGB model image. It is seen that when the database image is converted to HSV model (perceptually model), Figure 10.1a resembles Figure 10.1d. In HSV scene image, some query regions are also present in minor portions, other than the major areas. Using RGB model, the query image in Figure 10.1b is satisfactorily extracted from the database image as in Figure 10.1c. But when HSV model is used, the regions are extracted accurately as shown in Figure 10.1f as shown by the white portion in the region-extracted image.

Figure 10.2 is a multispectral scene of Landsat, SPOT, MOMS-02, and IKINOS, which offers spatial resolution between 30 and 40 m. The image covers an area of 204,152 km^2 and contains a mixture of urban, agriculture, and forestry areas around the world.

The database image in Figure 10.2a in HSV model is shown in Figure 10.2d. The region in the database image, equivalent to the query image in Figure 10.2c is actually the green colored region in the HSV-converted image in Figure 10.2d. So using HSV model, the exact portion is extracted as shown in Figure 10.2e which is equivalent to the query image (Figure 10.2f). In RGB, some unwanted regions (shown as white regions) are also extracted.

10.9 Fuzzy Multilayer Perceptron

Fuzzy Logic provides mathematical framework to capture a set defined by membership function expressing the grade of belongingness of each element in the set. The membership function lies between 0, that is, no membership to 1, i.e., full membership and larger values of membership denotes higher grade of belongingness. To take care of the inherent imprecision associated with many of our cognitive and decision-making processes such a fuzzy set theoretic framework has been widely applied.

The ANN are well known for their desirable characteristics of learning, adaptation, parallelism, generalization, and fault tolerance. A number of crisp interconnection networks have been proposed for a large number of applications including those related to the classification of remote sensing images. The advantages of such interconnection networks may be combined with the power of fuzzy sets in dealing with uncertainties to obtain better results.

In view of the above, we propose the design of such a set of fuzzy neural networks to satellite image pattern classification problems. In this chapter, we present the design of a fuzzy multiplayer perception network with back-propagation learning and fuzzy counter-propagation network. We also demonstrate the use of these networks for scene classification.

10.9.1 Fusion of Fuzzy Logic with Neural Networks

Fusion of fuzzy logic with neural networks have been suggested by a number of researchers [9–11,21–23,25]. Another method by Broomhead and Lowe [8], used radial basis function (RSF) network with radial function in a single layer network. RBF is a multidimensional function that depends on the distance between the input vector and the centre vector. It has an application in image processing, medical imaging etc. The advantages of fuzzy logic combined with the attractive properties of neural networks are bound to give better performance in the applications of fuzzy neural networks. Fuzzy neural systems in a number of industrial and control applications are on the increase [18,26].

In a fuzzy neural system, fuzzy systems and neural networks perform different tasks in a modular fashion, whereas in a fuzzy neural network, neuron functions, their input–output and the network parameters are fuzzified. Three broad areas in which fuzzy neural systems as well as fuzzy neural networks find application are: (a) modeling, (b) classification, and (c) prediction [3].

Fuzzy neural networks are gaining wide popularity due to their ability to handle fuzzy input–output relationships which are so often encountered in real life. The rules generated from the fuzzy neural networks can be applied to rule-based systems to overcome the bottleneck of rule extraction from experts. Thus neuro fuzzy networks have been found to be useful for generating rule base for expert systems.

10.9.2 Fuzzy MLP with Back-Propagation Learning

A fuzzy MLP system proposed by Mitra and Pal [25,26] uses fuzzy inputs represented in the form of quantitative, linguistic, or a combination of these. Further, the system can learn the input–output relationships and generate rules which can subsequently be used in a conventional expert system. The network was tested on vowel data, artificially generated patterns, and a medical diagnosis problem.

A logical version of MLP, suggested by the same authors [26] could learn the input–output fuzzy relationship in the form of if-then else rules. The output is presented with a certain amount of confidence level. The novel feature of the network is to query for missing information. The neurons are programmed to do min–max and product probabilistic sum operators represent the And-Or nodes at the hidden and output layers, respectively. Back-propagation algorithm is suitably modified for logical operations. The system can successfully handle very, more, less and not using Concentration (Con) and Dilation (Dil) operators.

Ishibuchi et al. [18] have also proposed a fuzzy back-propagation system with triangular inputs, outputs, and weights. A modified version of the same has been implemented here and successfully applied for supervised classification of RS images. The fuzzy triangular numbers at all stages were represented as fuzzy α-level sets.

Here we propose a scheme of representing the range of pixel gray values expressed as triangular fuzzy numbers. The output membership functions have also been represented here as triangular fuzzy numbers. Another interesting feature of this method is that it eliminates the dependency of training algorithm on the number of samples in the training set. Instead, the performance of the algorithm depends only upon the range of gray values occupied by the pattern classes in the image. Thus the training is carried out in a much faster manner as the input consists of one input vector of multi-band gray values represented by triangular numbers for each of the output classes. In this case, the number of input–output samples that is required to train the network is identical to the number of classes irrespective of number of labeled training pattern available.

For supervised learning, the output fuzzy membership values can also be derived from the training samples. The method of assigning fuzzy membership values to each class is based on M-distance and appropriate fuzzification process. The negative sample set of pixels belonging to the inappropriate classes can be eliminated during the fuzzy membership assignment scheme, thus facilitating a more efficient training of the MLP. By representing inputs, outputs, and weights as fuzzy triangular number, a complete fuzzy back-propagation scheme has been designed for classification of multispectral Indian Remote Sensing Satellite (IRS-IB) images. The results obtained have been observed to be more accurate than the crisp MLP with back-propagation learning.

10.9.3 Fuzzy Back-Propagation Classifier Architecture

The architecture of fuzzy MLP used in our scheme is basically similar to the crisp MLP. The structure of the three-layered network with single hidden layer can be generalized for multiple number of hidden layers. The number of nodes in the input layer is equal to the number of dimensions of the input pattern.

The gray value intensities of the pixels in each spectral band from the input to the network are in the form of normalized triangular values in the interval [0,1]. The formation of the triangular numbers is data dependent and will be discussed subsequently. Layers two and three are the hidden and output layers, respectively, of the conventional MLP with back-propagation learning. The nodes in layers two and three have fuzzy triangular weights and fuzzy biases.

The number of nodes in the output layer is equal to the number of output classes. The desired output pattern required for training the network can be in the form of binary values, positive real values, or triangular number as described later. An optional defuzzification layer has been added at the output of our system to map the fuzzy triangular output to a non-fuzzy (crisp) real number.

The neurons take the weighted sum of all the inputs and pass them through the sigmoid nonlinearities as in the crisp MLP network. During the learning session, the error signals flow in the reverse direction as in the case of conventional model and the real number arithmetic will be modified to interval arithmetic.

10.10 Fuzzy Counter-Propagation Network (CPN)

Kohonen's network in its basic form is an online version of the hard c-means algorithm. This algorithm can be generalized to become a fuzzy Kohonen's network to enhance its performance. The formulation is closely related to the fuzzy c-means algorithm suggested by Bezdek [3]. Fuzzy c-means algorithm is especially useful in applications where clusters touch or overlap. It is superior to its crisp counterpart in yielding more often the global optimum. Thus, the integration of neural network training with the fuzzy set concepts is expected to yield superior results in the area of pattern classification.

The goal of hard c-means algorithm as well as Kohonen's network is to cluster the input pattern into c clusters such that the total Euclidean distances between the input patterns and their nearest cluster centroids are minimized. The hard c-means algorithm can be fuzzified by making the concept of the Winner fuzzy. Thus the input pattern can evoke partial response in more than one node and the weights of more than one node can be updated proportional to the response. The learning takes place according to the win membership. The learning is effective in situations where the class boundary is fuzzy. The crisp algorithm can be modified as shown below:

1. Initialize the number of neurons with random weights. $W_i(0)$; $i=1,\ldots, N$.
2. Compute the distances with the input pattern x_k for all the competing neurons i. Distance measure $d(x_k,W_i)$ can be Euclidean or any other metric depending upon the data in hand.
3. Fuzzy computation involves computing win membership μ_{ki} of each neuron based on the distance measures $d(x_k,W_i)$
4. Fuzzy learning ensures that each neuron updates its weights according to the following rule:

$$W_j(t+1) = W_j(t) + \alpha(t)Z_i(t)\left[x_k - W_j(t)\right] \tag{10.1}$$

The fuzzy scaling function $Z_i(t)$ can be given by

$$Z_i = (\mu_{ki})^{f_m} \tag{10.2}$$

$$\mu_{ki} = \frac{\left(\frac{1}{D_{ki}}\right)^{1/(f_m-1)}}{\sum_{p=1}^{c}\left(\frac{1}{D_{kp}}\right)^{1/(f_m-1)}} \tag{10.3}$$

$$D_{ki} = d(x_k W_i) \tag{10.4}$$

when f_m is a real quality greater than 1 which is known as fuzzy generator. It follows that closer the neuron to the pattern, the larger its win membership μ_{ki} and the more it has to learn. In such learning, all neurons orient themselves toward the patterns of gradually decreasing levels proportional to the distance measure.

10.11 Fuzzy CPN for Classification of Remotely Sensed Data

10.11.1 General Description of the Test Scenes

The images obtained from the Linear Imaging Self Scanner-II (LISS-II) of the IRS-IB was used. The spatial resolution of each pixel is 36.25 m. Three images from three different study regions in the eastern part of India were chosen by the domain experts. These are four band multispectral images of 512×512 size. The domain experts from the Regional Remote Sensing Service Centre of the Eastern region were consulted for acquiring the training samples. Two images were containing 6 broad classes of land categories and the other image had 16 classes.

Data Set#1

The image known as data set#1 in our experiments (shown as false color composite (FCC) in Figure 10.3 (a)) consist of a coastal scene of Orissa, a state in India. It depicts the land/sea/river boundaries clearly. The sparse coastal

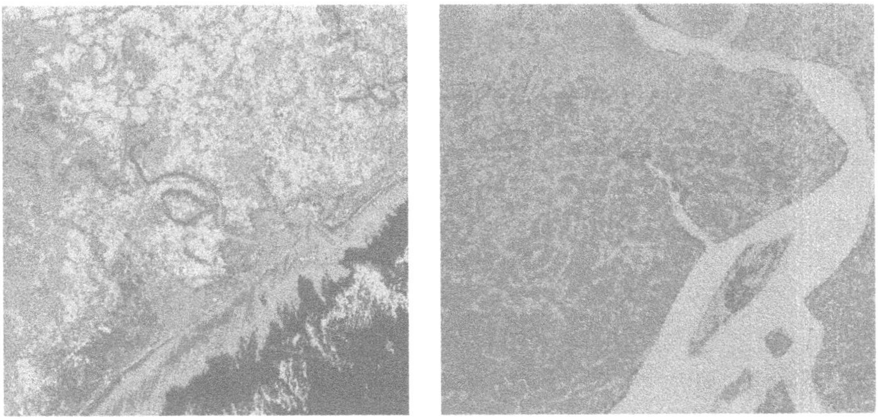

(a) (b)

FIGURE 10.3
(a) Image FCC, Scene 1 and (b) image FCC, Scene 2.

and inland vegetation can be seen in the land portion mostly covered by light colored alluvium (clay) soil. The terrain is flat and mostly barren. The meandering pattern of river gives rise to many cut of channels where the moisture content is perceived to be high. The beach contains a thin veneer of sand. The limes in front of the river mouth in the sea indicate high sedimentation zones. A detailed canal network used for irrigation can be easily visualized. Various classes contained in the image are listed below with their class numbers and the adopted color scheme.

1. Class 1: Midnight blue: Deep water 1
2. Class 2: Dark blue: Deep water 2
3. Class 3: White: Sand
4. Class 4: Light blue: Shore water 1
5. Class 5: Dark green: Sparse vegetation
6. Class 6: Light Green: Inland vegetation 1
7. Class 7: Dark Cyan: Shore water 2
8. Class 8: Parrot green: Inland vegetation 2
9. Class 9: Ink blue: River water 2
10. Class 10: Medium Cyan: River water 3
11. Class 11: Pink: Tanks
12. Class 12: Amber: Oxbow (vegetation)
13. Class 13: Light Cyan: Clear water
14. Class 14: Light magenta: Wet land 1
15. Class 15: Dark brown: Wet land 2
16. Class 16: Yellow: Dry land

In the above data set, four general classes and four subclasses have been identified for training of neural networks. The four general classes are, type of soil, vegetation, and freshwater bodies. The seawater could be visually identified by domain experts to contain three types of shades varying with the depth and shore water showing some turbidity. The land classes of wetland (two types), sand and dry land have been found to be containing overlapping classes and they are bound to produce errors of omission and commission. For example, sand and dry land are hard to distinguish. There are four subgroups of freshwater bodies which include clear water, lakes, and two types of river water.

Data Set#2

The image data set#2 consists of a navigable river namely Hooghly, depicting a harbor and a port in the interior of the river. FCC image is shown in Figure 10.3b. The surface of the land is made of mostly alluvium light gray clay and plain sand in some parts. The scene is mostly containing rich cultivable soil.

Streaks of vegetation can be seen along the sides of the canal network. This land is also flat. Large canal networks and river streams can be made out. A fallow tank, mudflats, and dense vegetation in the river island can be seen. The river shows turbid water. The following classes of objects are found in the scene.

The general color scheme employed is also listed.

1. Class 1: Light blue: Water
2. Class 2 Dark blue: Tidal flat
3. Class 3: Light green: Vegetation 1
4. Class 4: Dark green: Vegetation 2
5. Class 5: Red: Barren land
6. Class 6: Orange: Deep water

The images obtained from the LISS-II of the IRS-IB were used. The spatial resolution of each pixel is 36.25 m. Two images from two different study regions in the eastern part of India were chosen by the domain experts. These are four band multispectral images of 512×512 size. The domain experts from the Regional Remote Sensing Service Centre of the Eastern region were consulted for acquiring the training samples. Two images contained 6 broad classes of land categories and the other image had 16 classes.

The fuzzy CPN has been applied for classification of RS multispectral data. The fuzzy Kohonen layer consisting of 10×10 nodes was employed. Each node is having a weight vector of four elements corresponding to the four band inputs. The inputs have been normalized to the scale of 0 to 1. There are two learning constants α and β pertaining to Kohonen layer and Grossberg layer, respectively. The fuzzifier f_m can take real values greater than 1. It was observed that optimum results were available for values of f_m between 1 and 2. The best results were available for a value of 1.6.

The classified outputs of the scenes are presented in Figures 10.4a,b. The fuzzy CPN classified image outputs are much better in quality.

10.11.2 Experimental Results

Classification of multispectral RS data involves the assignment of a RS pixel with different spectral bands to a proper pattern class. The two images were used without any preprocessing, pruning, or statistical analysis before applying them for classification. It was observed that when the inputs were subjected to a normalization of the form given by the following equation, the training was further simplified.

$$x_i^{\text{norm}} = \frac{x_i - x_{\min}}{x_{\max} - x_{\min}} \tag{10.5}$$

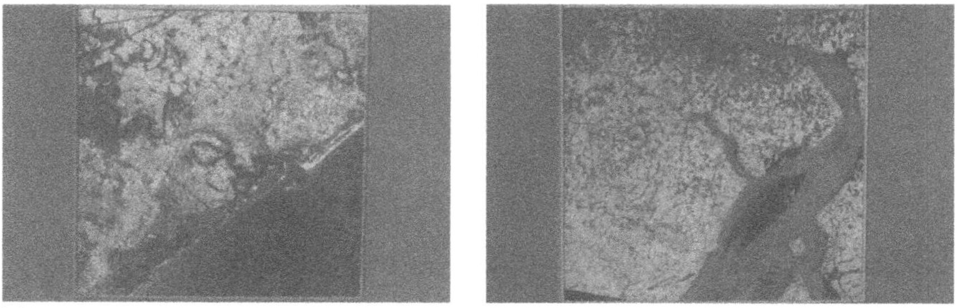

(a) (b)

FIGURE 10.4
(a) Image classified by Fuzzy Neural Network: Scene 1 and (b) image classified by Fuzzy Neural Network: Scene 2.

where

x_i is the input to be normalized

x_i^{norm} is the normalized input

$x_{\min}$ and $x_{\max}$ are the minimum and maximum values taken by the input in that dimension

In Scene 1, the coastal area vegetation has been brought out better. The linear features and the pockets of vegetation have been depicted clearly. Due to the fuzzy clustering nature of the algorithm, pixels belonging to a particular class tend to form compact clusters, giving an appearance of a segmented image. The algorithm performs with better generalization capability, compared to its crisp counterpart. Such classified outputs may be quite useful in image interpretation systems.

In Scene 2, the light vegetation along the canals has been clearly shown. The mind flats are misclassified as deeper water. Also the vegetation is misclassified as water which is clearly indicated in the error matrix of Scene 2. A large number of riverbank vegetation pixels have been misclassified as water pixels due to the shallow nature of ground in this region.

10.12 Summary

We have introduced the fundamentals of RS, and the techniques of segmentation, preprocessing, and classification. The concepts of fuzzy remote sensing and their advantages over their crisp counterparts have been elaborated. A number of fuzzy neural networks and their applications in classification of RS scene have been described.

References

1. Benediktsson, J.A., Swain, P.H., and Ersoy, O.K., Neural network approaches verses statistical methods in classification of multisourse remote sensing data, *IEEE Transactions on Geoscience and Remote Sensing*, 28, 540–552, 1990.
2. Bezdek, J.C., *Pattern Recognition with Fuzzy Objective Function Algorithms*, Plenum Press, New York, 1981.
3. Bezdek, J.C. et al., FC: Fuzzy c means clustering algorithm, *Computers and Geosciences*, 10(2–3), 191–203, 1984.
4. Bishof, H. et al., Multispectral classification of Landsat images using neural network, *IEEE. Transactions on Geoscience Remote Sensing*, 30(3), 846–855, 1992.
5. Bischof, H., Schneider, W., and Pinz, A.Z., Multispectral classification of landsat images using neural networks, *IEEE-GSRS*, 30, 482–490, 1992.
6. Biswas, P. and Majumdar, A.K., A multistage fuzzy classifier for recognition of hand printed characters, *IEEE-SMC*, 11, 1981.
7. Brooks, R.A., Symbolic reasoning among 3-d models and 2-d images, *AI*, 17, 285–348, 1981.
8. Broomhead, D.S. and Lowe, D., Multivariable functional interpolation and adaptive networks, *Complex Systems*, 2, 321–355, 1988.
9. Buckley, J.J. and Hayashi, Y., Fuzzy neural networks: A survey, *FSS*, 66, 1–13, 1994.
10. Cannon, R.L., Dave, J.V., Bezdek, J.C., and Trivedi, M.M., Segmentation of a thematic mapper image using the fuzzy c-means algorithm, *IEEE-GSRS*, 24(3), 400–408, 1986.
11. Carpernter, G.A. and Grossberg, S., Art 2: Self-organization of stable category recognition codes of analog input patterns, *AO*, 26(23), 4919–4930, 1987.
12. Foody, G.M. et al., Non-linear mixture modelling without end-members using an artificial neural network, *International Journal of Remote Sensing*, 18(4), 937–953, 1997.
13. Foody, G.M. and Arora, M.K., Incorporating mixed pixels in the training, allocation and testing stages of supervised classifications, *Pattern Recognition Letters*, 17(13), 1389–1398, 1996.
14. Foody, G.M. and Cox, D.P., Sub-pixel land cover composition estimation using a linear mixture model and fuzzy membership functions, *International Journal of Remote Sensing*, 15(3), 619–631, 1994.
15. Heermann, P. and Khazenie, N., Classification of multispectral remote sensing data using a back-propagation neural network, *IEEE Transactions on Geoscience Remote Sensing*, 30(1), 81–88, 1992.
16. Hepner, G., Logan, T., Rittner, N., and Bryant, N., Artificial neural network classification using a minimum training set: Comparison to conventional supervised classification, *Photogrammetric Engineering and Remote Sensing*, 56(4), 469–473, 1990.
17. Hunt, R.W.G., *Measuring Color: Ellis Horwood Series in Applied Science and Industrial Technology*, Halsted Press, New York, 1989.
18. Ishibuchi, H., Kwon, K., and Tanaka, H., A learning algorithm of fuzzy neural networks with triangular fuzzy weights, *Fuzzy Sets and Systems*, 71, 277–293, 1995.

19. Kelegama, D.C., Li-hua, L., and Jian-qin, L., Self organization map for clustering and classification in the ecology of agent organizations, *Journal of Central South University of Technology*, 7(1), 53–56, 2000.
20. Lee, J. et al., A neural-network approach to cloud classification, *IEEE Transactions on Geoscience Remote Sensing*, 28(5), 846–855, 1990.
21. Mannan, B. and Ray, A.K., Crisp and fuzzy competitive learning networks for supervised classification of multispectral IRS scenes, *International Journal of Remote Sensing*, 24(17), 3491–3502, 2003.
22. Mannan, B. and Ray, A.K., Fuzzy competitive learning network for classification of multispectral IRS scenes, in *Proceedings of the 55th National Conference on Communication*, pp. 300–305, Jan. 29–31, 1999.
23. Mannan, B., Roy, J., and Ray, A.K., Fuzzy ARTMAP supervised classification of multi-spectral remotely sensed images, *International Journal of Remote Sensing*, 19(4), 767–774, 1998.
24. Maselli, P.F. et al., Inclusion of prior probabilities derived from a non parametric process into the maximum likelihood classifier, *Photogrammetric Engineering and Remote Sensing*, 58, 201–207, 1992.
25. Mitra, S. and Pal, S.K., Fuzzy multilayer perceptron, inferencing and rule generation, *IEEE Transactions on Neural Network*, 6(1), 51–63, 1995.
26. Mitra, S. and Pal, S.K., Logical operation based fuzzy MLP for classification and rule generation, *Neural Networks*, 7, 353–373, 1994.
27. Page, G. and Gommm, B., Neurofuzzy techniques and applications (special issue), *FSS*, 79(1), 1996.
28. Paola, J.D. and Schowengerdt, R.A., A review and analysis of backpropagation neural networks for classification of remotely-sensed multi-spectral imagery, *International Journal of Remote Sensing*, 16, 3033–3058, 1995.
29. Palubinskas, G. et al., An evaluation of fuzzy texture based classification approaches for mapping regenerating tropical forest classes from Landsat TM, *International Journal of Remote Sensing*, 16, 747–759, 1995.
30. Smith, J.R., Integrated spatial and feature image systems: Retrieval, analysis and compression, PhD thesis, the Columbia University, New York, 1997.
31. Wang, F., Fuzzy supervised classification of remote sensing images, *IEEE Transactions on Geoscience Remote Sensing*, 28(2), 194–201, 1990.

11

MATLAB® Programs

11.1 Introduction

In this book, all the results which have been shown in different chapters have been implemented using MATLAB®. In many cases, the students find it difficult to implement their image processing algorithms in MATLAB. In view of this, we present in this chapter a brief note on MATLAB, followed by the common MATLAB commands, and finally the MATLAB codes of the various algorithms we have implemented.

MATLAB is a matrix laboratory; it was invented in the late 1970s by Cleve Moler. It is a numerical computing method for writing programs, which is created by MathWorks to provide an easy matrix or vector (a 1-D matrix) calculations, plotting of functions and data, implementation of algorithms, and interfacing with programs in other languages. To mention a few, it is also used in education such as in teaching linear algebra, numerical analysis, differential equation, and so on and is very much popular amongst scientists involved in image processing. MATLAB consists of many commands, which directly computes different matrix operations and other functions. Image is a 2-D matrix and thus using MATLAB, programming becomes very easy instead of writing in C. It performs intensive tasks faster than that of the traditional programming language such as C, C++, and FORTRAN. MATLAB also includes "for," "while" loops, and "if-else"commands as in C.

There are many examples in the MATLAB Tool Box. The best way to learn MATLAB is to work through some examples at the computer. MATLAB commands and help are available online.

MATLAB program and files always have filenames ending with "filename.m."

Some of the basic matrix commands, which are used in image processing, are as follows:

To display a matrix

$$a = [2 \quad 3; \quad 2 \quad 4]$$

$$a = \begin{matrix} 2 & 3 \\ 2 & 4 \end{matrix}$$

b=a.*a (element by element multiplication)

$$b = \begin{matrix} 4 & 9 \\ 4 & 16 \end{matrix}$$

Addition of two matrices – a+b
Inverse of a matrix – inv(a)
zeros(2, 3)

$$a = \begin{matrix} 0 & 0 & 0 \\ 0 & 0 & 0 \end{matrix}$$

ones (2, 3)

$$a = \begin{matrix} 1 & 1 & 1 \\ 1 & 1 & 1 \end{matrix}$$

Some of the MATLAB commands used in image processing are as follows:

1. Reading an image A.
   X=imread(A)
2. Displaying an image
   imshpw(A)
3. Histogram of an image A
   imhist(A)
4. Histogram equalization
   histeq(A)
5. Cropping an image – A small portion from the image is cropped.
   imcrop(A)
6. Rotating an image by 30°
   imrotate(A, 30)
7. Convert to unsigned 8-bit integer. The command (uint8) converts the matrix into an unsigned 8-bit integer. B can be any numeric object such as Double. The elements of an UINT8 range from 0 to 255. The result for any element of B outside this range is not defined and they are either assigned to 0 or 255. It is written as
   C=uint8(B)
8. Finding the edges of the image. There are many edge detection operators— Roberts, Sobel, Prewitt, Laplacian of Gaussian (LoG), etc. An example using Prewitt operator is as follows:

```
A=imread('rice.tif');
EIM=edge(I,'prewitt');
imshow(A), figure, imshow(EIM)
```

9. To convert an image to a binary image
   BIM=graythresh(A)
10. Converting a RGB color image to gray image
    rgb2gray(a)
11. Performing morphological operators on the binary image. There are many operations such as erode, dilate, thin, and open remove. An example of "thin" an image is shown.

```
A=imread('rice.tif');
imshow(A)
IM1=bwmorph(BW1,'thin');
figure, imshow(IM1)
```

11.2 MATLAB Examples

Few examples on fuzzy image thresholding, fuzzy region extraction, fuzzy color retrieval, and fuzzy edge detection are given that will show the implementation of fuzzy algorithms on image processing. At each step, the work performed by the part of the program is explained in "%" sign. This will help the readers to apply any fuzzy methods on image processing. In each example, we present the algorithm in a simple language *in italics* wherever necessary and then proceed to show the implementation using MATLAB.

Example 11.1

Consider an image. Apply fuzzy thresholding scheme to threshold an image that uses gamma membership function for finding the membership values of the pixels of an image and fuzzy divergence as a measure to find the optimum threshold.

```
clear all;
tic
a1=imread('cameraman.tif') % input image
dim=256;
a=imcrop(a1,[1 1 dim-1 dim-1]);
c=double(a);
[count,x]=imhist(a);
count;
```

```
y=dim*dim;
d=count/y;
d;
div _ vec _ init=[];
mx=max(max(c))+1;
mn=min(min(c))+1;
const=1/(max(max(c))-min(min(c)));
```

% finding the mean of the two regions%
(For finding the membership function for the object and back ground regions with respect to the means of the object and background regions, initially means are computed for the two regions)

```
for p=20:160% between the two peaks %
  s=0.0;
  for b=mn: p
  z=count(b)+s;
  s=z;
  z;
end
z;
s1=0.0;
for b=p+1:mx
  if p<mx
    z1=count(b)+s1;
    s1=z1;
    z1;
  else
 z1=0.0;
  end
z1;
end
z1;
t=0.0;
for b=mn:p
    m=count(b).*(x(b)-1)+t;
  t=m;
  end
m;
m1=fix(m*(1/z)); %mean of the object region
u=0.0;
for b=p+1:mx
  if p<mx
    n=count(b).*(x(b)-1)+u;
    u=n;
else
```

```
  n=0.0;
end
end
n;
n1=fix( (1/z1)*n);
```
%mean of the background region

% finding the gamma membership values of the pixels of the image using gamma membership function %

(gamma membership function is calculated using the means of the object and background to find the membership values of the pixels of an image)

```
for i=1:dim
  for j=1:dim
    if c(i,j) <p
      mem(i,j) =exp(-const*abs(c(i,j)-m1));
      mem(i,j);
    else
      mem(i,j) =exp(-const*abs(c(i,j)-n1));
      mem(i,j);
        end
    end
  end
u1=0.0;
```

% find the fuzzy divergence between an ideal thresholded image and the experimental thresholded image %

(For each threshold gray level, fuzzy divergence value is calculated between the ideal thresholded image and experimental thresholded image)

```
for i=1:dim
  for j=1:dim
  h= (2-(1-u(i,j) + 1).*exp(u(i,j)-1)-(1-1 + u(i,j)).*exp(1-
  u(i,j))) +u1;
  u1=h;
    end
  end
  div= (1/y)*h;
  div _ vec= [div _ vec _ init; div];
  div _ vec _ init=div _ vec;
  h1(p) =div;
  end
xval2=[];
```

% plot for the divergence value against gray level%

```
for xval=20:160
  xval1= [xval2; xval];
```

```
xval2 = xval1;
end
figure, plot(xval1, div _ vec);
xlabel('gray level')
ylabel('fuzzy measure');
hold off
```

% selecting the minimum fuzzy divergence value%
(The minimum divergence value is selected and the threshold corresponding to the minimum divergence value is selected as the optimum threshold.)

```
l = min(h1(20:160))
[p] = find(l == h1)
toc
```

Example 11.2

Consider a color query or a model image and a large scene image. Apply region extraction technique that extracts the regions in the scene image which are similar to the query or the model image provided. [*Hint: Region extraction uses fuzzy back projection algorithm*]

```
clear all;
a = imread('image.bmp'); % input image
tic
dim = 10; % query image size
dim1 = 120; % database image size
maxtr = [];maxtg = [];maxtb = [];
```

% calculating the maximum count for each channel of all the windows of the scene image%
(The model image is compared with the nonoverlapping windows of the scene image. The membership function is computed by normalizing the histogram of the query image and the histograms of the nonoverlapping windows in the scene image by a normalizing factor (norm). The normalizing factor is calculated by taking the maximum of the maximum counts of the histogram of each window of the scene image and the query image for each channel)

```
for i1 = 1:dim:dim1
  for i2 = 1:dim:dim1
  im = imcrop(a,[i2 i1 dim-1 dim-1]); %window of the scene image %
  imr = im(:,:,1); img = im(:,:,2); imb = im(:,:,3);
  % histogram for red channel%
  [count,x1r] = imhist(imr);
  countimr = count;
  % histogram for green channel%
  [count,x1g] = imhist(img);
```

```
   countimg = count;
   % histogram for blue channel%
   [count,x1b] = imhist(imb);
   countimb = count;
% maximum count of red channel in each window of the scene image%
   mx11r = max(countimr);
   maxt1r = [maxtr;mx11r];
   maxtr = maxt1r;
% maximum count of green channel in each window of the scene image%
   mx11g = max(countimg);
   maxt1g = [maxtg;mx11g];
   maxtg = maxt1g;
% maximum count of blue channel in each window of the scene image%
   mx11b = max(countimb);
   maxt1b = [maxtb;mx11b];
   maxtb = maxt1b;
  end
end
% maximum count among all the windows in each channel (normalizing
factor)%
mx2r = max(maxt1r);
mx2g = max(maxt1g);
mx2b = max(maxt1b);
% membership value calculation for the query image, c (query image) are
calculated by the normalizing factor%
c = imcrop(a,[1 1 dim-1 dim-1]);
cr = c(:,:,1);
cg = c(:,:,2);
cb = c(:,:,3);
[count,x2r] = imhist(cr);
countcr = count;
[count, x2g] = imhist(cg);
countcg = count;
[count, x2b] = imhist(cb);
countcb = count;
countc1r = (1/mx2r)*countcr;
countc1g = (1/mx2g)*countcg;
countc1b = (1/mx2b)*countcb;
convimr = zeros(dim1,dim1);
convimg = zeros(dim1,dim1);
convimb = zeros(dim1,dim1);
% membership value of red, green, and blue channels in the image
windows%
for i1 = 1:dim:dim1
  for i2 = 1:dim:dim1
```

```
  dr = [];dg = [];db = [];
  countdivr = 0.0;countdivg = 0.0;
  countdivb = 0.0;
  im = imcrop(a,[i2 i1 dim-1 dim-1]);
  imr = im(:,:,1);
  img = im(:,:,2);
  imb = im(:,:,3);
  im1r = double(imr);
  im1g = double(img);
  im1b = double(imb);
  [count,x1r] = imhist(imr);
  countimr = count;
  [count,x1g] = imhist(img);
  countimg = count;
  [count,x1b] = imhist(imb);
  countimb = count;
  countim1r = (1/mx2r)*countimr;
  countim1g = (1/mx2g)*countimg;
  countim1b = (1/mx2b)*countimb;
```

% computing the fuzzy ratio between the each database image and query image for each channel using fuzzy similarity measures%

```
for p = 1:256
  if countimr(p) > 0 |countcr(p) > 0
  counr = (min(countc1r(p),countim1r(p)))./
  (min(countc1r(p),countim1r(p)) +
  0.5*min(countc1r(p),1-countim1r(p)) + 0.5*min(countim1r(p),1-
  countc1r(p)));
    countdivr = [dr;counr];
    dr = countdivr;
  else
   countdivr = 0;
    countdivr = [dr;countdivr];
    dr = countdivr;
   end
if countimg(p) > 0 |countcg(p) > 0
  coung = (min(countc1g(p),countim1g(p)))./
  (min(countc1g(p),countim1g(p)) +
  0.5*min(countc1g(p),1-countim1g(p)) + 0.5*min(countim1g(p),1-
  countc1g(p)));
    Countdivg = [dg;coung];
    dg = countdivg;
  else
    countdivg = 0;
    countdivg = [dg;countdivg];
    dg = countdivg;
```

```
  end
if countimb(p) >0 | countcb(p) >0
counb = (min(countc1b(p),countim1b(p)))./
(min(countc1b(p),countim1b(p)) + 0.5*min(countc1b(p),1-
countim1b(p)) +0.5*min(countim1b(p),1-countc1b(p)));
  countdivb= [db;counb];
    db=countdivb;
  else
    countdivb=0;
    countdivb= [db;countdivb];
    db=countdivb;
  end
  end
hr= (countdivr);
hg= (countdivg);
hb= (countdivb);
```

% back projection of fuzzy ratio on to the image%
(The fuzzy ratio is back projected on to the image i.e., image values are replaced by the fuzzy ratio to obtain a matrix)

```
for i=i1:i1+ (dim-1)
  for j=i2:i2+ (dim-1)
    sr=im1r(i-i1+1,j-i2+1);
    convimr(i,j) =hr(sr+1);
    sg=im1g(i-i1+1,j-i2+1);
    convimg(i,j) =hg(sg+1);
    sb=im1b(i-i1+1,j-i2+1);
    convimb(i,j) =hb(sb+1);
      end
    end
  end
end
convimr; convimg; convimb;
tempr=max(max(convimr));
tempg=max(max(convimg));
tempb=max(max(convimb));
finalimr= (1/tempr)*255*convimr;finalimr1=uint8(finalimr);
finalimg= (1/tempg)*255*convimg;finalimg1=uint8(finalimg);
finalimb= (1/tempb)*255*convimb;finalimb1=uint8(finalimb);
[count,x9r] =imhist(finalimr1);
```

% finding the highest peak for each channel%

```
countxyr=count;
mxconr=max(countxyr);
[x9r] =find(mxconr== countxyr);
[count,x9g] =imhist(finalimg1);
```

```
countxyg = count;
mxcong = max(countxyg);
[x9g] = find(mxcong = = countxyg);
[count,x9b] = imhist(finalimb1);
countxyb = count;
mxconb = max(countxyb);
[x9b] = find(mxconb = = countxyb);
```

% **thresholding at the highest peak for each channel%**
(*The image is thresholded at the highest peak to obtain a region-extracted image*)

```
for w1 = 1:dim1
  for w2 = 1:dim1
    if finalimr1(w1,w2) = =  (x9r-1)
      segimr(w1,w2) = 0;
    else segimr(w1,w2) = 1;
    end
  end
end
for w1 = 1:dim1
  for w2 = 1:dim1
    if finalimg1(w1,w2) = =  (x9g-1)
      segimg(w1,w2) = 0;
    else segimg(w1,w2) = 1;
    end
  end
end
for w1 = 1:dim1
  for w2 = 1:dim1
    if finalimb1(w1,w2) = =  (x9b-1)
      segimb(w1,w2) = 0;
    else segimb(w1,w2) = 1;
    end
  end
end
figure,  imshow(b)
figure,  imshow(c)
image1 = segimr + segimg + segimb;
tmp = max(max(image1));
image = (1/tmp)*image1;
[count,x10] = imhist(image);
countim = count;
mxc = max(countim);
[x10] = find(mxc = =  countim);
for w3 = 1:dim1
  for w4 = 1:dim1
```

```
    if image(w3,w4) == (x10-1)/255
      finim(w3,w4) = 0;
    else finim(w3,w4) = 1;
    end
  end
end
% filtering with Gaussian filter%
h = fspecial('gaussian',3,4);
finalimage = filter2(h,finim);
figure, imshow(finalimage)
toc
```

Example 11.3

Consider a color image database. Select a query image from the image database. Apply a fuzzy method for color image retrieval that uses fuzzy histogram and fuzzy measure to select the top 10 images that are almost similar to the query image.

```
clear all;
tic
```

% list the set of images in the flag image database (99 images)%
list={'flag.bmp'};

% constant calculation for membership function%
Calculating the maximum and minimum values of the histogram counts of each of the images in the database for each channel

```
for k1 = 1:99
  a = imread(list{k1});
im = imcrop(a,[1 1 127 127]);
imr = im(:,:,1);
img = im(:,:,2);
imb = im(:,:,3);
bin = 32;
[count,x1] = imhist(imr,bin);
countimr1 = count;
[count,x2] = imhist(img,bin);
countimg1 = count;
[count,x3] = imhist(imb,bin);
countimb1 = count;
mximr(k1) = max(countimr1);
mximg(k1) = max(countimg1);
mximb(k1) = max(countimb1);
mnimr(k1) = min(countimr1);
```

```
mnimg(k1) =min(countimg1);
mnimb(k1) =min(countimb1);
end
```

Calculating the maximum value and minimum value among all maximum and minimum histogram counts for each image for each channel

```
mxr=max(mximr(:));
mxg=max(mximg(:));
mxb=max(mximb(:));
mnr=min(mnimr(:));
mng=min(mnimg(:));
mnb=min(mnimb(:));
% constant calculation
constr=1/(mxr-mnr); constg=1/(mxg-mng); constb=1/(mxb-mnb);
% computing the histograms for the query image and each of the images in the database%
s= [];
for k1=1:99
  data _ im=imread(list{k1});
  im=imcrop(data _ im,[30 30 127 127]);
  bin=32;
  b=imread('query.bmp');
  ref=imcrop(b,[30 30 127 127]);
refr=ref(:,:,1);
refg=ref(:,:,2);
refb=ref(:,:,3);
k=ones(bin,1);
% histogram of the query image%
  [count,x11] =imhist(refr,bin);
  countrefr=count;
  [count,x12] =imhist(refg,bin);
  countrefg=count;
  [count,x13] =imhist(refb,bin);
  countrefb=count;
% histogram of the images in the database%
  imr=im(:,:,1);
  img=im(:,:,2);
  imb=im(:,:,3);
  [count,x1] =imhist(imr,bin);
  countimr=count;
  [count,y1] =I mhist(img,bin);
  countimg=count;
  [count,z1] =imhist(imb,bin);
  countimb=count;
% gamma membership values of each database image using the constant%
  memrefr=exp(-constr*abs(countrefr-countimr));
```

```
  memrefr;
  memrefg=exp(-constg*abs(countrefg-countimg));
  memrefg;
  memrefb=exp(-constb*abs(countrefb-countimb));
  memrefb;
% gamma membership values of the query image%
  memimr=exp(-abs(countimr-countimr));
  memimr;
  memimg=exp(-abs(countimr-countimr));
  memimg;
  memimb=exp(-abs(countimr-countimr));
  memimb;
% find the similarity between the each database image and query image using fuzzy similarity measures for finding a close match with the query image%
tar=1-(sum(abs(memrefr-memimr))./sum(memrefr+memimr));
tag=1-(sum(abs(memrefg-memimg))./sum(memrefg+memimg));
tab=1-(sum(abs(memrefb-memimb))./sum(memrefb+memimb));
sim=[s; tar+tag+tab];
s=sim;
end
sim;
% sorting the images in ascending order according to similarity%
[u,v] =sort(sim);
[u,v]
toc
```

Example 11.4

Consider an image. Using template-based method for fuzzy edge detection, find the edge-detected image. The template shows the possible direction of the edges.

```
clear all;
tic
  a11=imread('image.bmp');
  a2=dyaddown(a11,'m',1);
  dim=100; row=dim; col=dim;
  a=0.3; b=0.8;
  image=imcrop(a2,[1 1 dim-1 dim-1]);
% a set of 16 fuzzy edge templates that shows the possible directions of the edges of an image%
m1= [a,a,a;0,0,0;b,b,b]; m2= [a,a,b;a,b,0;b,0,0];
```

```
m3 = [b,b,b;0,0,0;a,a,a]; m4 = [b,a,a;0,b,a;0,0,b];
m5 = [b,a,0;b,a,0;b,a,0]; m6 = [a,0,b;a,0,b;a,0,b];
m7 = [0,0,0;b,b,b;a,a,a]; m8 = [0,b,a;0,b,a;0,b,a];
m9 = [a,a,a;b,b,b;0,0,0]; m10 = [a,b,0;a,b,0;a,b,0];
m11 = [0,0,0;a,a,a;b,b,b]; m12 = [0,a,b;0,a,b;0,a,b];
m13 = [b,b,b;a,a,a;0,0,0]; m14 = [b,0,a;b,0,a;b,0,a];
m15 = [b,0,0;a,b,0;a,a,b]; m16 = [0,0,b;0,b,a;b,a,a];
b1 = double(image);
mx = max(max(b1));
img = b1/mx;
edgim = zeros(dim);
dim1 = dim + 1;
edgimage = zeros(dim1);
```

program to consider border points of the image

```
r1 = img(2,:); r2 = img(row-1,:); c1 = img(:,2);
c2 = img(:,col-1); bor1 = [0,r1,0]; bor2 = [0,r2,0];
p1 = [c1,img,c2]; f2 = [bor1;p1;bor2];
f2(1,1) = img(1,1); f2(row + 2,col + 2) = img(row,col);
f2(1,col + 2) = img(1,col); f2(row + 2,1) = img(row,1);
```

% Computing the max-min value for each position

a. Finding a similarity between each of the image templates and the image window to know in which direction the edges present and computing the maximum of the divergence values between the template and image window%

```
for i = 2:dim1
for j = 2:dim1
im = [f2(i-1,j-1),f2(i,j-1),f2(i + 1,j-1);
f2(i-1,j),f2(i,j),f2(i + 1,j); f2(i-1,j + 1),f2(i,j + 1),f2(i + 1,j + 1)];
ed1 = 2-(1-im + m1).*exp(im-m1)-(1-m1 + im).* exp(m1-im);sum1 = min
(min(ed1));
ed2 = 2-(1-im + m2).*exp(im-m2)-(1-m2 + im).*exp(m2-im);
sum2 = min(min(ed2));
ed3 = 2-(1-im + m3).*exp(im-m3)-(1-m3 + im).*exp(m3-im);
sum3 = min(min(ed3));
ed4 = 2-(1-im + m4).*exp(im-m4)-(1-m4 + im).*exp(m4-im);
sum4 = min(min(ed4));
ed5 = 2-(1-im + m5).*exp(im-m5)-(1-m5 + im).*exp(m5-im);
sum5 = min(min(ed5));
ed6 = 2-(1-im + m6).*exp(im-m6)-(1-m6 + im).*exp(m6-im);
sum6 = min(min(ed6));
ed7 = 2-(1-im + m7).*exp(im-m7)-(1-m7 + im).*exp(m7-im);
sum7 = min(min(ed7));
```

```
ed8 = 2-(1-im + m8).*exp(im-m8)-(1-m8 + im).*exp(m8-im);
sum8=min(min(ed8));
ed9 = 2-(1-im + m9).*exp(im-m9)-(1-m9 + im).*exp(m9-im);
sum9=min(min(ed9));
ed10=2-(1-im+m10).*exp(im-m10)-(1-m10+im).*
  exp(m10-im); sum10=min(min(ed10));
ed11=2-(1-im+m11).*exp(im-m11)-(1-m11+im).*
  exp(m11-im); sum11=min(min(ed11));
ed12=2-(1-im+m12).*exp(im-m12)-(1-m12+im).*
  exp(m12-im);s um12=min(min(ed12));
ed13=2-(1-im+m13).*exp(im-m13)-(1-m13+im).*
  exp(m13-im); sum13=min(min(ed13));
ed14=2-(1-im+m14).*exp(im-m14)-(1-m14+im).*
  exp(m14-im); sum14=min(min(ed14));
ed15=2-(1-im+m15).*exp(im-m15)-(1-m15+im).*
  exp(m15-im); sum15=min(min(ed15));
ed16=2-(1-im+m16).*exp(im-m16)-(1-m16+im).*
  exp(m16-im); sum16=min(min(ed16));
```

b. Selecting the minimum divergence among the 16 divergence values and is positioned at the center of the template position

```
sum _ tot=[sum1,sum2,sum15,sum16,sum3,sum4,sum5,sum6,sum7,su
m8,sum9,sum10,...
  sum11,sum12, sum13, sum14];
  edgim(i-1,j-1)=max(sum _ tot);
  end
end
```

% forming a divergence matrix by calculating the max–min divergence for all pixel positions%

```
edgim; temp=max(max(edgim));
maxim= (1/temp)*(edgim);
sd=uint8(255*maxim);
figure, imshow(image);
```

% threshold the divergence matrix to obtain an edge-detected image%

```
for i1=1:dim
  for j1=1:dim
    if maxim(i1,j1)>0.23(threshold value)
    w2(i1,j1)=1;
    else w2(i1,j1)=0;
    end
  end
  end
  figure, imshow(maxim); w3=bwmorph(w2,'thin'); figure,
  imshow(w3);
  toc
```

Example 11.5

Select a low contrasted image. Using INT operator, obtain an enhanced image. Apply any membership function to fuzzify the image.

```
tic
image = imread('bloodcell2.bmp');
a = rgb2gray(image);
a12 = dyaddown(image,'m',1)
dim = 126;
c = imcrop(a12,[10 1 dim-1 dim-1]);
d = double(c);
mx = max(max(d)); mn = min(min(d));
% membership function calculation%
Img = (d-mn)./(mx-mn);
% applying INT operator%
for i = 1:dim
  for j = 1:dim
  if img(i,j) <= 0.5
    enh(i,j) = 2*img(i,j)^2;
  else
    enh(i,j) = 1-2*(1-img(i,j))^2;
  end
  end
end
enhanceimage = (L-1)*enh;
figure, imshow(uint8(enhanceimage))
figure, imshow(c)
```

Problems

1. Define fuzzy set. Explain with examples.
2. What are the different types of fuzzy membership function? Plot the fuzzy membership function of the following:
   1. Pixel brightness
      a. Very bright
      b. Bright
      c. Neutral
      d. Dark
      e. Very dark
   2. Homogeneity
      a. Very small
      b. Small
      c. Medium
      d. Large
      e. Very large
3. Apply fuzzy hedges on any image selecting any arbitrary membership function and find the effect on the final image.
4. Consider a fuzzy set:

$$A = \{1/0.0, 2/0.2, 3/0.3, 4/0.6, 6/0.5, 7/0.8, 9/0.9, 10/1.0\}$$

   Find (i) core, (ii) support, and (iii) 0.3 cut of the fuzzy set.
5. Consider the following two fuzzy sets:

$$A = ((x_1/0.4), (x_2/0.9), (x_3/0.8), (x_4, 1))$$

$$B = ((x_1/0.2), (x_2/0.5), (x_3/0.7), (x_{4,}, 0.9))$$

   Find $A \cup B$, $A \cdot B$, $A - B$

6. Consider two fuzzy relations:

$\Re_1$	y_1	y_2
x_1	0.2	0.4
x_2	0.5	0.7
x_3	0.1	0.3
x_4	0.9	1.0

$\Re_2$	y_1	y_2
x_1	0.3	0.9
x_2	0.7	0.5
x_3	0.9	0.3
x_4	0.8	1.0

Determine: $\mu_{\Re_1 \cdot \Re_2}(x, y)$, $\mu_{\Re_1 + \Re_2}(x, y)$

7. For the following relation, check for the max–min transitivity.

$\Re$	x_1	x_2	x_3
x_1	0.8	0.6	0.4
x_2	0.4	1	0.1
x_3	0.3	0.5	0.9

8. Define measures of fuzziness and fuzzy similarity measures.
9. What are the properties of the measures of fuzziness and fuzzy similarity measures?
10. Explain the use of fuzzy image preprocessing.
11. Consider an image. Apply INT operator and NINT operator on an image and find out the difference in the enhanced image.
12. Consider a low contrast medical image. Apply 3×3 Gaussian filter to smooth the image. Then use any of the fuzzy enhancement algorithms to enhance the image contrast.
13. Apply all the four fuzzy enhancement algorithms on an image and find which of the algorithm gives better result. Compare the result with histogram equalization.
14. Repeat Problem 12 for local image enhancement. [*Hint:* Divide the image into windows and apply enhancement algorithm to each window.]
15. Select a noisy image. Apply $m \times m$ fuzzy median filter and test for the "m" values for 3,5,7. Repeat the same by using Gaussian fuzzy filter with median value.

16. Consider a noisy image and apply weighted fuzzy mean filter and fuzzy weighted mean filter.
17. What are the different types of thresholding.
18. Write a program that computes the histogram of an image. Consider two bi-level images containing some objects. Histogram of one image is narrow and deep and the other is shallow and wide.
    1. Threshold the image using fuzzy divergence and fuzzy compactness.
    2. Calculate the ground truth threshold value.
    3. Find the segmentation error when compared with the ground-truth image.
19. Create an image IM1 containing geometrical objects such as squares, circles, and rectangles with constant gray level against a constant gray level different background. Apply Gaussian noise with a given standard deviation and create a new image IM2. Threshold the image using any fuzzy thresholding method that will threshold the image clearly.
20. Create an image of printed or handwritten text. Determine the threshold value of an image that uses exponential membership function and linear indices of fuzziness as a measure of fuzziness.
21. Consider a pathological image of blood vessels. Apply suitable method that thresholds the image. [*Hint: you may apply filter to remove any noise present in the image.*]
22. Consider a bi-level image with two peaks. Find the minimum value between the two maxima and threshold the image using the minimum value. Does this lead to the best segmentation?
23. Repeat Problem 20 using S-membership function.
24. Define match-based region extraction.
25. Develop a black-and-white scene image with multiple regions. Select any small region from the scene image which is a query image. Query image may be present at many places in the scene image. Design a match-based region extraction procedure that will successfully segment the query image from the scene image.
26. Repeat Problem 24, by rotating the scene image by 40°. Query image will be selected from the rotated scene image.
27. Select a colored (red, green, and blue) scene image which can be grasslands, mountains, rivers, etc. As in Problem 24, select a small region (query image) and develop a match-based region extraction procedure that will successfully segment the query regions completely.

28. Design a set of fuzzy rules for edge detection and implement fuzzy reasoning method to detect the edges of an image using these fuzzy rules.
29. Using fuzzy method that considers edge templates and fuzzy divergence as a measure of fuzziness, design an edge detection technique. Show the results for different threshold values and select the threshold value at which the edges are clearly detected. Also see the effect of the edge detected results when the values of the fuzzy edge templates are changed.
30. Compute the edge detected image using fuzzy template method and compare this method with the conventional edge detection schemes, such as using Sobel, Prewitt operators. Also show the result for different fuzzy edge templates.
31. Repeat the above method using color (RGB) image. *Hint:* Detection is done for each channel.
32. Select an image and blur the image using averaging filter. Apply fuzzy edge detection method to detect the edges of the image.
33. What are the different types of color spaces? Explain with the help of diagrams.
34. Choose an image database and select a query image from the database. Apply global fuzzy histogram technique to find the top 20 images in the database that almost resemble the query image. Calculate the precision/recall percentages.
35. Repeat Problem 34 using local fuzzy histogram method. *Hint*: Local histogram is the histogram of the windows of the image.
36. Choose an image database and select a query image that is not present in the database. Apply local histogram technique to find the closest match of the query image to any of the images in the database.
37. Repeat Problem 35 using HSV and CIELab color space. Calculate the precision/recall percentages for both the color spaces and notice the difference with the RGB image.
38. Choose an x-ray image of any human organ. Segment the image using any fuzzy clustering method into several homogeneous regions.
39. In a two-class problem, suggest an algorithm for supervised classification of fuzzy patterns using a set of labeled fuzzy patterns and compare its performance with the Bayes' classifier.
40. What are the techniques of fuzzification and defuzzification of patterns? Design a fuzzy multilayered perceptron neural network with error back-propagation and investigate its utility with respect to crisp non-fuzzy multilayered perceptron techniques.

41. What is Fuzzy Counter-propagation network? How do you cluster the pattern samples using Fuzzy Kohonen's network?
42. Why are fuzzy techniques preferred over their non-fuzzy counterpart for segmentation and classification of remotely sensed images?
43. Design a content-based image retrieval system when a query image is presented for retrieval from a large image database.

Index

G

H

For Product Safety Concerns and Information please contact our EU representative GPSR@taylorandfrancis.com
Taylor & Francis Verlag GmbH, Kaufingerstraße 24, 80331 München, Germany

www.ingramcontent.com/pod-product-compliance
Ingram Content Group UK Ltd.
Pitfield, Milton Keynes, MK11 3LW, UK
UKHW020159200926
475702UK00004B/538
* 9 7 8 1 4 3 9 8 0 7 0 8 8 *